8-95

SOLVENT ABUSE: A Guide for the Carer

A Guide for the Carer

JOHN S. CAMERON

CROOM HELM
London & Sydney

© 1988 John S. Cameron
Croom Helm Ltd, Provident House,
Burrell Row, Beckenham, Kent BR3 1AT
Croom Helm Australia, 44–50 Waterloo Road,
North Ryde, 2113, New South Wales

British Library Cataloguing in Publication Data

Cameron, John S.
  Solvent abuse: a guide for the carer.
  1. Solvent abuse   2. Youth——Substance use
  I. Title
  362.2′93   HV5822.S65
  ISBN 0–7099–4862–X

Printed and bound in Great Britain by Mackays of Chatham Ltd, Kent

Contents

Acknowledgements

Solvent abuse is a topic encompassing many fields of interest; so many people and organisations have contact with this subject, whether from the clinical aspect, or other relevant spheres.

I would like to express my gratitude and thanks to the many people who have helped make this book possible: The Greater Glasgow Health Board for their support, not only for this book, but for their assistance in the running of our solvent abuse clinic, in particular: Mrs M. Hay (retired), Miss M. Aitken, Mr T. Goodsir, Mr P. Paterson, Dr G. Forwell, Mr P. Murray; and the Solvent Abuse Team: Dr P. Misra, Jack Murphy, Mike Tucker, Anne Riddell, Adrian Murtagh, Bill Underwood, Carol Moran, Yolanda Hunter, Frank Green, Stephen Watt, Fiona Kane, Grace May, Dr Danny Daniels, Liz McGregor, and of course Janette Pollock who painstakingly corrected and typed the manuscript, not forgetting Betty Diamond.

Dr Brian Wells of Strathclyde University for his kind use of information from *Psychedelic Drugs* (Penguin), Mr M. Shapiro from the Institute for the Study of Drug Dependence for allowing me to use their information on legal issues (Chapter 5), Dr H.E. Akerman (Evode Ltd) for the use of his paper 'The Constitution of Adhesives and its Relationship to Solvent Abuse', Mrs J. Whitwell (British Aerosols Manufacturers Association) for allowing me use of the information in Dr J. Roberts' article 'Abuse of Aerosol Products by Inhalation' in Chapter 3 (p. 16) (both papers published by *Human Toxicology* in 1982).

Thanks also to:

The British Adhesives Manufacturers Association.

Ron Hanvey for his information on Children's Panels in Scotland, Alistair Ramsay (Staff Tutor, Education Department, Strathclyde) for his drug education package (Drugwise 12–14-year-olds, Strathclyde Education Department), and the Scottish Drug Forum (W. Slavin).

The Library Staff at the College of Nursing (Glasgow Royal Infirmary).

Dr P. Misra for his article on hypnosis.

I am indebted to the staff at St George's Hospital Medical

School for their kind permission to use their mortality statistics (Appendix A); *The British Medical Journal*; Mr A. Vanet of the Crown Office, Edinburgh, the Home Office provided information (reproduced with kind permission of the Controller of Her Majesty's Stationery Office) pages 5 and 103–4 Appendix C and the Department of the Environment; Strathclyde Police (Chief Superintendent M. Frood), Metropolitan Police Crime Prevention Service, Strathclyde Regional Council, Strathclyde Social Work Department (Addictions); Occupational Therapy Department (Gartloch and Acorn Street Hospitals), Imogen Clark and Lesley Wilkie; Addiction Information and Resource Unit (Glasgow); Jennifer and Katherine Tsang; Mr Barrie Liss and Re-Solv for their kind assistance and information.

I sincerely hope that this book will not only help to answer some of the questions brought up by the subject of solvent abuse, but will encourage the setting up of more, badly needed centres for young people who are abusing solvents.

Note: Throughout the book drugs such as cannabis, LSD, amphetamines and magic mushrooms are mentioned. It would be a mistake to take solvent abuse on its own, as recent trends show that there is considerable mixing of these different substances; also there are marked similarities between solvents and magic mushrooms, so this area can be expanded to include the misuse of hallucinogenic drugs, LSD and cannabis.

Figures and Tables

Part I

Background to Solvent Abuse

1

Historical Background to Substance and Drug Abuse

INTRODUCTION

The adage that 'nothing' is new under the sun' can certainly be applied to the field of substance abuse. Although modern laboratory-produced drugs and aerosols were not available centuries ago, equally addictive substances were, and still are, available for man to experiment with.

SUBSTANCE USE IN THE PAST

Vapour or gas inhalation, for the purpose of intoxication or to communicate with the spirit world, has been with us since time began. In ancient Greece, a young priestess was credited with experiencing visions after inhaling vapours from rock fissures.

Considerable experimentation with a variety of substances has been going on since earliest man, not only for their soporific effect, but as a food source, and through the ages a collective knowledge has formed of which substances constitute sustenance, and which can provide escapism. This knowledge has been inherited from generation to generation. Different cultures have adopted specific practices, especially from religious ceremonies where mind-altering substances would be used, for example, in burning spices, incenses, etc. to obtain visions.

Historically, some facts may actually be legends, and there is doubt as to whether the substances mentioned are the same in use today; however very similar substances would definitely have been used (Wells, 1973). For example, cannabis is said to have appeared in the pharmacopoeia of the Emperor Shen

3

Nung in 2737 BC. Though a Shen Nung may well have ruled at about that time, and although cannabis has been known to exist in China for many millennia, this apparently precise reference is based on legend rather than clear historical sources.

Cannabis use, of course, has a very ancient medicinal tradition throughout the eastern world, and it has also been used for spiritual inspiration, identified in fact by many scholars as the 'heavenly guide' referred to in the Hindu Atharva, and Rig-Vedas.

The humble cactus (*Lophophora williamsii*), which projects only 7.5 cm above barren soil, produces in its fleshy top one of the strangest drugs in the pharmacologist's collection. It was used by both pre-Aztecs and Aztecs for its psychedelic properties (de Ropp, 1957), by slicing it into discs which were dried and then become peyote or mescal 'buttons' (Wells, 1973).

Opium was in use as an aid to medicine more than 1800 years ago; more recently, it was used in the 19th century in Cambridge, England, not only for medicinal purposes (apparently the Fens were endemic with malaria), but also taken socially — for example, one custom was to drink beer with a piece of opium dissolved in it. This custom became so well established that brewers and publicans produced opium beer in order to meet the popular taste (Gossop, 1982).

Alcohol has been present with us throughout history. Records of the uses and abuses of alcohol have been found in ancient Egypt. Plato, in ancient Greece, felt it was barbaric to drink undiluted wine, and in the book of Genesis (IX) in the Old Testament, Noah is noted for his drinking of wine (Gossop, 1982).

Coca leaves, derived from the coca bush, are found mainly on the eastern slopes of the Andes (Peru and Bolivia), and have been used by the local inhabitants for centuries (incidentally, the Indians only receive a fraction of cocaine from chewing the leaves that a cocaine user would from buying cocaine on the street). Chewing the leaves helps alleviate the hard, cruel life they often have to endure. Coca-cola, in fact, actually had cocaine as one of its ingredients until 1903.

Heroin, which belongs to the opiate family, was first produced in the late 19th century, and was sold on the market as a cough remedy. At the beginning of this century, opium, morphine and heroin consumption began to show an increase.

DRUG ABUSE OVER THE PAST HUNDRED YEARS

Then in the late 1940s and early 1950s drug abuse started to become a major problem, the most popular drugs of abuse at this time being:

(1) Cannabis (this could actually be obtained by prescription up to 1973).

(2) Amphetamine (a stimulant); there is a history of abuse during both the Second World War and Vietnam war, but it was eventually controlled under misuse of drugs legislation in 1964 (Institute for the Study of Drug Dependence, 1984).

(3) Lysergic acid diethylamide (LSD), derived from ergot (a fungus found growing wild on rye, etc.), was first produced by Stoll and Hoffman in 1938, but it was not until Hoffman accidentally inhaled some of this material in his laboratory in 1943, and found himself caught up in a full-scale psychedelic experience, that the psychological effects of the drug were realised (Wells, 1973).

'One indication of the changing pattern of drug misuse is the number of addicts who are formally notified to the Home Office by doctors.

The number of addicts notified in the United Kingdom in 1983 (5850) was 42 per cent up on 1982; and in 1984 the number had increased to 26 per cent over the 1983 figure (7400). But this is probably only a small proportion of the actual number of chronic misusers. Many will not have sought medical treatment and will not therefore have been notified. Research carried out in 1981 in two urban areas in England suggested that the number of notified addicts was a five-fold underestimate of opioid addicts in the local population at that time, and that there may have been a similar number of people misusing other drugs.

That there may be considerable local variations is suggested by a study carried out by Ditton and Speirits (1981) in a Scottish city which found that only one in ten heroin users had been notified.' (Misuse of Drugs. Notification of and Supply to Addicts Regulations, 1973 SI, 1973 No. 799).

Experimentation with substances other than specific drugs for their possible effect have also been increasing, albeit on a smaller scale. In the last century, for example, 'laughing gas'

5

(nitrous oxide) was extremely popular for the euphoric feeling it produced. Ether and chloroform were also sniffed for their enjoyable effects.

When gas lamps were used to illuminate stairways etc., it was not unusual to find them being blown out and the gas being bubbled through a soft drink, for consumption. For a number of years, petrol has also been used for sniffing purposes. This practice is not just confined to Europe, but also in sparsely populated areas where perhaps there is not such a selection of other substances available, for instance, in the outback of Australia, and in parts of Africa.

Solvent abuse as we know it today has its roots in the United States (glue being the main substance), probably since the 1940s. In the United Kingdom, apart from sporadic episodes, the problem has only become noticeable since the early 1960s.

DEFINING SOLVENTS

It would be extremely difficult to differentiate between substances, in order to claim that one substance, for example butane gas, is to be used for *solvent* abuse. If we then take magic mushrooms in this context, what do we call taking these substances — drug-taking, perhaps, because any mushroom containing psilocin and psilocybin would come under the Misuse of Drugs Act.

Volatile substance abuse (VSA) has also been used to describe the taking of volatile substances (such as aerosols). Polydrug abuse (more than one), and most certainly poly-solvent abuse, is not uncommon, and the mixing of solvents and drugs is becoming more popular and widespread. Therefore, the field of abuse in general can be an extremely complex and at times 'sophisticated' minefield of different substances, effects and methods of achieving a 'high'.

There are some 'misleading' terms surrounding the substance abuse spectrum, although unintentionally. For the sake of simplicity, I would call substances such as glue, butane gas, aerosols, perfumes, cleaning agents, etc., that are taken for their potential effects, solvent abuse or volatile substance abuse. Cannabis, LSD, magic mushrooms etc. come under the heading of drugs. It is not so long ago that the term 'glue-sniffing' was used to describe all varieties of substance abuse;

certainly glue was then very popular, but the term disguised other forms of substance abuse that were also prevalent.

SOLVENT AND DRUG MIXING

Over the past few years there appears to be an upsurge in the level of activity associated with the mixing of both solvent and drugs, in order to achieve a heightened effect. The most commonly used mixtures are:

(a) cannabis and butane gas/aerosol spray;
(b) amphetamines (speed) and butane gas/aerosol spray;
(c) LSD and butane gas/aerosol spray.

In discussion with users of these 'mixed substances and drugs', the answers to why they indulge in what appears to be entirely separate cultural substance or drug misuse are quite varied and interesting:

(1) The majority of users claim that they get a far quicker and pleasanter experience than just using one substance or drug alone.
(2) It is cheaper, because they are not actually using so much of either substance and because of sharing.
(3) Some users have claimed to achieve 'psychedelic dreams' complete with flashing colours, and 'happy hallucinations' (this only applies to the LSD and gas/aerosol users). Therefore one is apt to feel puzzled, especially when these particular users claim that taking either LSD or butane gas/ aerosol on its own does not give such a good effect. Because of the small number actually abusing this mixture, and possibly the users' credibility, it is difficult to assess the effects, and how often the experience is attempted fully.
(4) It is more acceptable to their peers than simply inhaling butane gas/aerosols.

Certainly the mixing of solvents with cannabis or amphe- tamines is becoming more popular; possibly 80–90 per cent of *all substance and drug mixing* use this mixture. The remainder tend to use LSD with the appropriate substance. Recently, a noticeable trend has been use of butane gas/aerosols instead of the particular drug (generally cannabis or amphetamines and occasionally LSD). General reasons for this are:

(1) Financial — it is cheaper to buy butane gas/aerosols than drugs.
(2) Availability — sometimes drugs can be difficult to obtain (police action perhaps), whereas butane gas/aerosols can be bought or stolen from a variety of sources.
(3) The user has been in trouble with the courts over drugs already and feels that he or she is safer having a can of gas on his/her person instead of drugs.
(4) The user prefers the 'effect' from solvents to that from drugs.

2

Addiction and How to Detect It

INTRODUCTION

The word 'addiction' is synonymous with alcohol, solvent and drug misuse. However, it is a misleading term as different substances produce different patterns of use. The term 'dependence', which can be subdivided into two parts, physical and psychological, is more appropriate as it allows us to understand more fully the different problems encountered when dealing with such a wide issue as substance abuse.

DEFINITIONS

Physical dependence

An example of physical dependence would be the amount of physical discomfort or distress caused by the withholding of a substance. For example, if an individual has been taking alcohol for some considerable time, and the alcohol is then withheld, various symptoms such as trembling, convulsions, delirium tremens (DTs) can occur. So when an individual has an intense dependence on a substance there is a tendency to increase the amount being consumed to prevent any possible withdrawal effects (see Chapter 6 for further discussion).

Psychological dependence

This is recognised as being the most important and widespread kind of dependence. As with physical dependence, psychological dependence can vary with different substances. A person who has been taking an opiate (heroin) drug on a long-term basis can develop strong psychological dependence on this drug, yet the use of LSD, for instance, rarely leads to psychological dependence.

It should be borne in mind, however, that an individual's psychological make-up must be taken into account; also some users will exist in a subculture removed from their original environment (see Chapter 6, for further discussion).

Tolerance

This is quite simply the ability of the body to take larger amounts of whatever drug he is on to maintain the same effect. If an inexperienced solvent abuser were to inhale the same amount of solvent as a user of several years standing, the result would be nausea/vomiting, disorientation, unconsciousness, possibly even death. With the use of the opiates, fatal overdoses can occur when the user takes his usual dose after a break during which tolerance has faded, for example, after serving a prison sentence.

SOLVENT ABUSE

Butane gas, aerosols, glue and any other volatile substance can be included under this heading.

Physical dependence

There is very little evidence to show that physical dependence is a problem here; however, a small percentage of users report aching limbs, stomach, back and headaches, anxiety and sleeplessness.

Psychological dependence

A number of young people involved with heavy use of solvents, who have emotional or family problems or are immature, will develop psychological dependence.

Recreational and experimental users do not as a rule develop this type of dependence.

Tolerance

This develops very quickly with solvents. Regular sessions of solvent-taking can build up tolerance within a few months. For example, at the end of 1986 an older glue sniffer who claimed to be spending almost £25 a day on glue was referred to a Glasgow clinic. Subsequent investigations into this almost unbelievable figure showed that there now appears to be a very small minority of users spending this amount of money on solvents. A byproduct of this type of solvent-taking is crime — money is needed to finance this sort of intake — or actual theft of substance.

MAGIC MUSHROOMS

Physical dependence

There is no evidence to indicate that the use of magic mushrooms will lead to physical dependence.

Psychological dependence

This will be present in certain cases. Unfortunately there are no studies available which give an assessment of the effects of extended, frequent use (Institute for the Study of Drug Dependence, 1984).

Tolerance

This will develop rapidly; however, with intervals of abstinence the user can prevent the tolerance level rising too quickly.

11

CANNABIS

Physical dependence

There would appear to be mild symptoms of physical dependence, tiredness and listlessness among them.

Psychological dependence

Regular users will have psychological dependence.

Tolerance (reverse tolerance)

There are varying opinions on this. The experienced cannabis user will take *less* than a newcomer to the drug, possibly due to a lower initial anxiety about the effects, and obviously more experience with its use. This is contrary to the usual course of drug abuse, where the user becomes 'tolerant' to the drug's effect and requires progressively *more* to achieve the same state. Some researchers claim tolerance increases moderately.

AMYL NITRITE

Physical dependence

This substance can cause some physical dependence — tremors, nausea, vomiting, headaches.

Psychological dependence

When this occurs the usual symptoms include anxiety and depression.

LYSERGIC ACID DIETHYLAMIDE (LSD)

Physical dependence

No physical dependence has been noticed.

Psychological dependence

This is very rare; only a small percentage of users have become psychologically dependent.

Tolerance

This builds up very quickly, within a few days. LSD has a low toxicity in man, in that it can be difficult to actually overdose on it.

POLYDRUG USE

The subject of solvent abuse cannot be taken in isolation. Recent trends are pointing to misuse of solvents and drugs together, and to polydrug use. Some young users are becoming increasingly sophisticated in how they obtain a high: butane gas and magic mushrooms, or a mixture of gas and either cannabis or pills (DF118 or amphetamine) have been noted.

In addition, some drug users have moved to gas or aerosols, either for financial reasons, or a lack of drugs, police action, etc. Some users also claim that:

(a) there are not so many health problems!
(b) aerosols cannot be adulterated.

The worker in the field must be aware of solvent and drug interaction, and the terms used, and when dealing with users must be able to demonstrate that he is knowledgeable about what is being abused, and how. In drug abuse, a large subculture has sprung up, with new terms for drugs/substances and related activities (see Glossary, p. 142). With solvent abuse, however, this has not happened to any great degree; very few street terms exist for the substances that are used, the main reason probably being that most solvent users are only transitional — they will sniff a substance for a period of time, and then for various reasons cease, leaving only the chronic sniffer, who forms only a small percentage of users. It could be that there is more 'charisma' in taking drugs than solvents.

HOW TO DETECT SOLVENT ABUSE

This is probably the most popular question that many people including parents and professionals ask: 'How can I tell if someone is abusing solvents/substances?' The short answer is: it is very difficult to state with conviction if a young person is indeed taking solvents. Many agencies have put out a very creditable checklist of the signs and symptoms of an abuser, which unfortunately could apply to a large percentage of adolescents. Commonsense must prevail when making a decision as to whether or not a young person is abusing solvents. The main criteria could be:

(1) A change in sleep patterns: going to bed very late, and loath to get up in the morning.
(2) Cheeky, truculent, moody, non-communicative to parents or teachers.
(3) Spots around the mouth/nose (consistent with the holding of a bag of glue against the mouth and nose). However, in butane gas sniffing, spots and blemishes do not appear. To add a further complication, most adolescents have spots anyway!
(4) Playing truant from school. An important symptom of solvent abuse, this will lead the youngster into further trouble, from both the school authorities, and outside influences. He will be away from supervision for several hours, he will need money. Together with wintry rainy weather these all add up to extra problems that the truant will face, and complicate the original sniffing problem.
(5) Overnight disappearance. This can be extremely distressing for parents, particularly if the person is very young. It does happen that solvent parties/groups go on later than the participants realised, so they stay out all night, not wanting to go home still under the influence.
(6) Loss of appetite, and possibly weight. Taking solvents does suppress the appetite, and mealtimes are forgotten if they are being taken on a regular basis.
(7) 'Hangover' type of symptom.
(8) Slurred speech, loss of coordination, memory lapses.
(9) Hallucinations.

The most telling symptoms that would unarguably prove that a youngster is involved in solvent taking would be:

(a) definite disappearance of perfume, butane gas, glues from the home;

(b) empty solvent containers hidden in the bedroom or lying outside the window;

(c) remains of solvents on bedclothes, handkerchiefs, etc., smell of solvents in the bedroom, or indeed in the home where there is no reasonable explanation.

(d) if a definite smell of solvents can be detected on the breath;

(e) glue stains on the clothes, particularly anoraks and coats.

Schools can also play a part in the detection of solvent abuse of young people, by paying regard to:

(a) persistent truancy;

(b) dreamlike state or having hallucinations;

(c) unruly behaviour;

(d) lack of attention in class;

(e) unusual laughter;

(f) fellow pupils calling others 'glue head' or 'sniffer', etc.;

(g) frequent use of handkerchiefs in class;

(h) continually sniffing/sucking shirt sleeves/jacket cuffs;

(i) smell of solvents in the classroom.

It would be ideal that if one discovers a young person who has been indulging in solvent-taking, one could sit down and discuss the problem, but unfortunately, it does not happen like this. The abuser generally denies involvement at first, and in a number of cases the parents will want to believe him. Young people can give very convincing reasons why they are not sniffing, but if a parent or teacher, or indeed anyone, has definite suspicions, approaching a treatment centre for advice and help can be the first step, as the latter are well used to offering advice.

3

Substances: Sources, Methods of Abuse, and Effects

INTRODUCTION

In the 1960s solvent abuse was commonly known as 'glue-sniffing', with good reason, as glue was generally the main substance to be abused. Nowadays, use is more sophisticated, as many other substances are now being taken for their suitable effects.

This chapter gives a comprehensive list of all the known substances (there are possibly many others being abused of which the author is unaware), including magic mushrooms, nutmeg, banana skins, etc. In addition cannabis is included because not only is it a very popular drug, but it can be taken in a variety of ways for its alluring, slightly hallucinogenic properties. LSD (what it is and what it does) is also included as there are marked similarities between it and magic mushrooms.

GLUE

The main proprietary brands are the favoured types. Glue itself has slid down the popularity stakes as other substances are learned about and passed on. Certainly there is always a large core of young people who swear by the effects of glue and do not try any other substance. Glue is usually the first substance that is abused, and it is known as a 'good standby' when other more volatile substances are unavailable.

16

Sources

In spite of extensive advertising, etc. there are still shops which will, either unwillingly or knowingly, sell young people glue (see Chapter 5). If this source is unavailable, glue can be found or stolen from many places — homes, garden sheds, shops or building sites.

The more commonly used solvents, found in adhesives, and capable of being abused, are shown in Table 3.1. Some are highly flammable, some non-flammable, but all are narcotic on inhalation.

Table 3.1: Solvents Employed in Adhesives

Hydrocarbons	Esters	Ketones	Chlorinated hydrocarbons
Toluene	Ethylacetate	Acetone	Methylene chloride (dichloromethane)
Hexanes and heptanes	Isopropyl acetate	Methyl ethyl ketone (MEK)	1,1,1-Trichloro-ethane (also known as Genklene)
Alicyclic hydro-carbons of similar volatility to n-hexane			

Source: Akerman (1982)

Mixed solvent blends are often used in adhesives, to ensure the desired flow, evaporation rate, solvency, solids content, and other technical properties. It would be very unfair to blame adhesive manufacturers supplying a substance that can be so easily abused; indeed, the British Adhesives Manufacturers Association have taken an extremely responsible position issuing posters, etc. to wholesalers, and assisting various groups who help solvent abusers.

Evode, for example sent a poster (Table 3.2 below) to its wholesalers, and through them to its retailers. Note that it is solely for employees of the retailer. The company are not in favour of poster campaigns directed at the general public, feeling that this simply tells people what type of product to misuse, and where it can be obtained.

'As a general point of interest, the adhesive manufacturer is often asked 'why don't you put something in your adhesive to stop the sniffer?' This possibility has been fairly thoroughly

Table 3.2: Evode Poster

Staff Notice — Glue-Sniffing
The Facts
1. 'Glue-sniffing' is the name given to the practice of inhaling solvents from products, for the intoxicating effects derived from such misuse.
2. Children between the ages of 11 and 17 often experiment with glue-sniffing, sometimes on a group basis. Comparatively few become habitual sniffers as this abuse is not habit forming.
3. The intoxicating effects can lead to damage to physical or mental health.

The Products
4. Solvent-containing products that may be misused in this way include some adhesives, dry cleaning susbtances, antifreeze, nail polish remover, paint stripper, shoe cleaners and dyes.

The Law
5. Glue-sniffing is not a criminal offence. You may, however, refuse to sell solvent-containing products to children whom you suspect of misusing them in this way.

Signs to Look For
1. Children who purchase solvent-containing products more frequently than you would expect.
2. Children who appear to have pooled their pocket money to buy such products.
3. Children with red eyes or heightened colouring, or other signs of intoxication such as slurred speech.

A Responsible Approach to the Problem
If you have reason to suspect that a minor is seeking to purchase products for sniffing, report the matter to your manager.
If the circumstances are serious he may tell you not to sell the product, and decide to report the matter to parents, school teachers, social workers or police officers.
Poster issued in the interest of public health for the notice of the employees of the retailer.

studied over many years on both sides of the Atlantic, but for various reasons it has been found to be impracticable. As an example of what was generally found, we may consider the case of the use of allyl isothiocyanate, otherwise called 'volatile oil of mustard'. This material was added to an adhesive in sufficient quantity so that it would deter most sniffers. Then it was found to suffer from the following faults:

(1) It is toxic, and prolonged contact may cause vesication and slow-healing ulcers (Merck Index, 1960). Inhalation of the vapour may cause asthma, watery eyes and sneezing, and local contact may give rise to contact dermatitis (Sax, 1979).

Its addition to an adhesive would therefore present problems to the manufacturer that would be very expensive to overcome, particularly in the light of changes made in health and safety at work legislation since the use of such additives was first proposed.

(2) It reduces the bond strength of the adhesive.

(3) It discourages the purchase and use of the material for legitimate use.

(4) If it is used for legitimate purposes, it represents an as yet unknown risk to the health of the user.

(5) The cost of including the additive would be passed on to the legitimate user, with no corresponding benefit to him.

(6) As a reactive chemical, it has been shown to degrade in the adhesive, so that in a period much less than the shelf-life of the adhesive, the odour, and hence the deterrent effect, disappears.

(7) It has been found that it is *not* aversive to all sniffers.

(8) Those not averse to it and thus misusing it are at *greater* risk because of its toxicity than they would be with untreated adhesive.

(9) Where the additive is effective, the sniffer turns to some other solvent-containing material, as yet unadulterated, but possibly more hazardous.

The generally held view on aversive additives in this country was adequately summarised by a government minister in his speech in a parliamentary debate (Hansard, 1980). The years of expensive research that would be involved in looking for a suitable additive for just one commercial product, say an adhesive, would not be justified, because even if a research project did yield such an additive capable of deterring the sniffer, the same additive would be unlikely to be suitable for use in the alternative solvent-containing product, say a hair lacquer, to which the sniffer might then have recourse.

Proper research into the causes as to why people are induced to inhale solvent vapour, followed by treatment of these causes, is a better project for the expenditure of available resources.' (Akerman).

Method of abuse

Glue is generally poured into a crisp bag, put up to the nose and mouth and breathed in. To enhance this effect a plastic bag is put over the head; the abuser will then breathe in and out, obviously a very dangerous practice, as not only fumes from the adhesive are inhaled, but lack of oxygen can cause unconsciousness leading to death, if not discovered in time. Occasionally 'gang rituals' consist of an initiate having not only to sniff a bag of glue, but also to eat the hardened glue in it.

Effects

With inhalation, the vapours are absorbed through the lungs into the blood system, and then into the body organs. The length of time taken before the user feels 'high' obviously depends on many factors — age, weight, whether any food has recently been taken, length of glue sniffing experience, and mood. Generally 3–5 minutes is sufficient for the user to begin to feel high, then disinhibited, and at this stage a number of users have reported feeling aggressive. However, this stage is soon passed (it should be emphasised that it can be dangerous to approach anyone taking solvents at this time; it is far better to wait until some time has elapsed). Hallucinations can be experienced. I have heard some users describing frightening ones usually concerned with monsters, etc., while others have described seeing Superman and Batman.

One youngster's bizarre hallucinations have stuck in my mind. He was a ten-year-old boy who had been sniffing glue for approximately six months. He used to experience being one of Snow White's dwarfs; he would sniff glue every day, and each one of the dwarfs would get a turn in his hallucination, Dozy one day, Sleepy the next and so on. Fortunately he eventually began to take an interest in football and glue was soon forgotten.

The final stage is similar to effects from an alcohol binge with slurred speech, uncoordination, nausea, vomiting, slowed respirations, headaches; drowsiness and unconsciousness can also occur. Chronic users, however, generally stop before they reach the final stage, wait a short interval then start again. It is the inexperienced users who usually go on to the final stage.

AEROSOLS

Aerosols when used normally are perfectly safe; only when the vapours are artificially and deliberately concentrated can problems occur. The British Aerosol Manufacturers Association (like the British Adhesives Manufacturers Association) is also extremely concerned with misuse of its products. The Association believes that education, in the fullest sense of the word, is required to help young people who experiment with sniffing. Aerosols are devices containing a built-in energy source that allows the product to be released in the form of mist, spray or foam at the touch of a button. There are over 200 different types of aerosol on the UK market (Aerosol Review, 1981), and a guide to their history, science and manufacture has been published by the British Aerosol Manufacturers Association (1975). The benefit of using aerosols compared with conventionally packaged products has also been evaluated and documented by the Association (1980).

Most aerosols on the market use liquified gases as the energy source, but dissolved gas systems are also used. The amount of liquified gas in a product can be as little as 5 per cent (as in shaving foam), to more than 90 per cent (as in a pain relief spray). Figure 3.1 illustrates the working of a typical aerosol.

Figure 3.1: Cross-section of an Aerosol

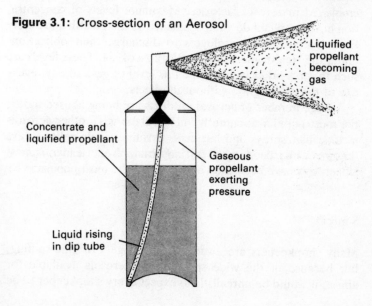

Liquified propellant becoming gas

Concentrate and liquified propellant

Gaseous propellant exerting pressure

Liquid rising in dip tube

Cosmetic and toiletry aerosols usually contain alcohol, and are propelled by halogenated hydrocarbons or a blend of these with hydrocarbons. Household products sometimes include petroleum distillates and are usually propelled by hydrocarbons. Industrial aerosols contain a wide variety of solvents including petroleum distillates and chlorinated types, the usual propellants being, again, halogenated hydrocarbons. A detailed analysis of propellant usage was published in 1978 (British Aerosol Manufacturers Association, 1978).

The principal ingredients in an aerosol product are the active ingredients, a solvent system and the liquified gases. The solvents include water (as in a starch product), alcohol (as in a hairspray or perfume), and kerosine (as in insecticide spray). Many other types of solvents are used, particularly in the industrial market.

The liquified gases are of two principal types: halogenated hydrocarbons, mainly dichlorodifluoromethane (propellant 12); and trichlorofluoromethane (propellant 11). The second type consists of deodorised butane which contains some propane in order to adjust its vapour pressure. Aerosols today often contain a mixture of both types of gas. Both have a low order of toxicity and have been used safely for a considerable period of time. Both have been subjected to intensive toxicological scrutiny to ensure safety to the aerosol user and to people employed in aerosol factories. Maximum levels of concentration have been laid down for guidance by the Health and Safety Executive to allow employers to discharge their obligations under the Health and Safety at Work Act 1974. These levels are such that a worker can breathe this level of gas, all day, every day of his working life without ill-effects.

A large number of aerosol products are being abused today, the most popular apparently being butane gas[1]; other aerosols include hair sprays and lacquers, perfumes, pain relief sprays (known as PR), deodorants, and industrial solvent cleaner. Almost all aerosols have been tried but these are the most popular.

Sources

Many shopkeepers are aware of the dangers of glue-sniffing, but because of the wide spectrum of aerosols available for abuse, it would be unrealistic to expect every shopkeeper to be

22

aware of what they are, and not to sell them to young people. In the home, there are usually any number of aerosols available for abuse, and unless a parent is suspicious the aerosols can be sniffed for a considerable time before action is taken.

The British Aerosol Manufacturers Association has a history of regular consultation with government departments to discuss current issues since it was set up in 1960. When news reached the Association of sniffing deaths in the United States, a meeting was set up with the Home Office and their medical advisers from the DHSS in 1968. The problem was far from simple, however, and ideas to introduce a lachrymator into aerosols to deter would-be sniffers were quickly agreed to be impracticable, particularly as toiletry products were often involved.

The meeting also fully recognised that sniffing was clearly abuse of the product by only a small sector of the population, and it would be inequitable to consider withdrawal of safe and effective products from the market for this reason. In addition, the meeting examined label warnings similar to those in the United States to see whether this would be a sufficient deterrent. The Association felt strongly that warnings used in the United States had actually given wider currency to the abuse system and even indicated those aerosols most suitable for use!

The earlier widescale press coverage followed by the educational campaign had also effectively publicised the possibility of using aerosols to obtain altered states of consciousness. The potential of early death did not seem to have the required deterrent effect among youngsters in the United States. Perhaps they were even motivated by bravado and rebellion against the authorities. After examining all the facts, the meeting decided that there should be no warning labelling as in the United States, and that the media should be encouraged to give as little publicity as possible to those incidents which did occur. The Home Office also consulted with their drug abuse advisory committees throughout the country to see whether this 'low key' policy was acceptable to them, and they unanimously agreed that the least possible publicity in this area was the best course of action, a policy pursued by both industry and government departments. Mortality statistics show that this policy was the correct one when viewed against the United States experience.

The British Aerosol Manufacturers Association has been involved in a number of consultations with the Home Office and the Consumer Safety Unit of the Department of Trade since then, but this agreed policy is still valid.

Method of abuse

There are varying methods of inhaling these aerosols: they can be sprayed onto a handkerchief, sleeve or rag, held up to the nose and sniffed; or, using a small room (for example a bathroom), blocking any vents (keyholes, etc.) and releasing two or three cans into the air, one can simply sit there and breathe in the fumes. A more dangerous method sometimes employed is to put a plastic bag over the head and release a can inside the bag, and it is not unknown for some users to spray under their bedclothes while in bed.

Effects

These can be felt within 2–3 minutes, and consist of elation, 'high feeling', drowsiness, hallucinations, mood swings. After 15–20 minutes, nausea, vomiting and headaches are extremely common.

Butane gas

Small tubes of gas refills now appear to be out of fashion, having been replaced by large cans. A nozzle on top of the can is pressed against the teeth and a jet of gas projected to the back of the throat. Unfortunately plastic bags have been introduced to this already very dangerous mode of abuse, thus ensuring that even more gas is inhaled by putting the bag over the head and releasing the gas into the already enclosed space.

There are some interesting trends amongst butane gas users that are currently quite distinct from other substances. From a study of almost 600 abusers the following points were noted.

(1) Butane gas (and other aerosols) were more popular among girls than taking glue — in fact only a small

percentage of girls actually took glue.

(2) Butane gas, far more than any other solvents, is used in 'mixing cocktails', for example, cannabis and butane gas, or amphetamine and butane gas (LSD and butane gas has also been tried by older users; it has been reported (Wells, 1973) that flashbacks of previous hallucinogenic trips can be achieved by inhaling deodorant sprays, but whether the users were aware of this is unknown) (Kramer and Pierpaoli, 1971).

(3) Unlike most glue-sniffing sessions which last for approximately 45 minutes, butane gas sessions can, and do, last for several hours and longer. One case that comes to mind: five years ago a young girl was referred to the Acorn Street Clinic, Glasgow for her habit of abusing gas. In spite of rigorous intervention, positive social skills assistance, etc. she was never able to stop completely, although she would abstain for up to three months at a time. She attended the Clinic periodically. Her method of abusing butane gas was to buy ten or twelve large cans at a time, then stay in her room for up to 72 hours, sniffing the gas and sleeping. She would rarely eat or drink during this period. At the time of writing, these sessions were down to one every 4 months. This case is certainly an extreme one, but it serves to illustrate the lengths to which some chronic abusers will go to achieve their own particular high.

Effects

Taking into account once again age, weight, etc., butane gas users reach a high quicker than with any other substance. An almost immediate effect has been described; certainly within 1½ minutes the effects can be felt, hallucinations being the main one with butane gas. Other symptoms include nausea, dizziness, mood swings, disinhibition, headaches, drowsiness and unconsciousness.

Butane gas hallucinations can be extremely vivid, and very few users have not experienced these. However, it must be remembered that there is a degree of expectation among users particularly in a group setting, where they will discuss their hallucinations quite openly with each other, and the level of 'expectation' will be quite high.

25

LYSERGIC ACID DIETHYLAMIDE (LSD)

First produced synthetically in 1938 by Stoll and Hoffman, it is derived from ergot (a fungus found growing wild on rye, etc). Microscopic amounts of the chemical only are required to produce the psychedelic effects for which LSD is so famous.

The drug has a history of being used in the treatment of individuals, as an aid in psychotherapy, and also for the treatment of alcohol problems. One of its properties seems to be that it reduces tension and dissolves the defences and reserves normally concerned with maintaining the social face we present (Wells, 1973).

LSD began to have a cult following particularly in the 1960s and early 1970s; the hippie movement was a great exponent of the mystical and religious effects that LSD could give.

Sources

It is usually obtained illegally, but it can be made quite simply by a competent chemist.

Method of abuse

Although usually available in tablet form (tablets of acid), it can also be found impregnated into stamps (generally with a cartoon figure on them), or in ampoule form (the laboratory type as used for experimentation), and occasionally in sugar cubes. Very few users of LSD inject the drug.

Effects

The entire experience (trip) can take approximately 12 hours. The first stage consists of rather an anxious half-hour, which can be accompanied by nausea; then comes the 'psychedelic effect' — mood changes, bright lights, excitement, preoccupation with any object that is noticed, heightened self-awareness, and sometimes religious feelings are experienced. Bad trips can occur, more so in a person who is unstable or has an existing psychiatric condition. After the main 'psychedelic' effect has worn off, there is tiredness and sometimes slight depression. Some users experience 'flashbacks', a reoccurrence of their LSD experience, for anything up to a year or so after taking the LSD.

MAGIC MUSHROOMS

Throughout time numerous varieties of hallucinogenic plants have been used by man for their effects. The ancient Aztecs worshipped cactus that gave them peyote (the cactus top is used for psychedelic purposes by slicing it into discs which are then dried and become peyote or mescal buttons; see Wells, 1973). Psilocybe is a mushroom also used by Aztecs for their religious ceremonies, producing euphoria and psychedelic effects.

In the United Kingdom possibly only a dozen hallucinogenic plants are used for these effects. The main ones pertinent to this book are the:

(a) liberty cap (*Psilocybe semilanceata*),
(b) fly agaric (*Amanita muscaria*).

Liberty cap can be taken in fairly high doses as it is relatively non-toxic to man (amounts used vary tremendously from six to 30 mushrooms.) This type of mushroom is probably the most popular, is available in most areas of the United Kingdom, and can easily be mistaken for other types of mushrooms. It is generally 6–8 cm high with a light brown cap.

Fly agaric has a history of being taken by certain Siberian tribes (Metzner, 1971), who not only ingest the mushroom, but also the urine passed by themselves and others who have taken it. It is white-spotted and red-topped, and can be 10–20 cm high.

Over the past few years the popularity of mushrooms as a substance of abuse appears to have waned. This may possibly be due to several factors:

(a) Many other substances available for abuse are widely known.
(b) There may be seasonal unavailability.
(c) There are bad side-effects: vomiting, nausea, sweating, cardiac disturbances.

27

(d) There is a lack of knowledge of how to differentiate between mushrooms.

Method of abuse

Generally, mushrooms are either eaten or made into 'mushroom tea'. A pot is put over a fire and water and mushrooms are mixed together and then drunk. Others will mix this solution with tea, and it is not unknown for alcohol also to be mixed with 'mushroom tea'.

(Peyote can either be chewed and swallowed, or made into tea. The taste and smell are apparently foul and vomiting and nausea are common; it can also be injected.)

Effects of *Psilocybe semilanceata*

It should be strongly emphasised that taking mushrooms for their effect is *extremely dangerous*. Some mushrooms, while capable of causing hallucinogenic effects, are poisonous, and others may have to be 'cooked' properly before they are relatively 'safe' to be taken.

Effects usually occur after 20–40 minutes and, depending on how much is consumed, can last 4–8 hours. Nausea and vomiting are frequent. Generally the user will fall asleep, but euphoria and hallucinations can also be present. The effects of psilocybin can last up to 12 hours. Symptoms can also include ataxia, impaired cognitive control of thought and attention, but with increased visual and auditory sensitivity (Isbell, 1959).

The active ingredients of *fly agaric* are muscarine, atropine and bufotenine (the latter being chemically very similar to psilocybin and DMT (dimethyltriptamine) and thus presumed to play some psychedelic role (Wells, 1973)).

Effects of fly agaric

Sweating, nausea, reduced cardiac rate, and other unpleasant digestive and vegetative changes are produced. If taken in sufficient quantity, it will cause the heart to cease completely and the user will die (Wakefield, 1958). Mood changes include euphoria, feelings of omnipotence, and also perhaps powerful feelings of persecution and aggression, and visual and perceptual changes.

CANNABIS

Cannabis comes from the female hemp plant, *Cannabis sativa*, which contains numerous chemicals; delta-9-tetrahydro-cannabinol (THC) is the main one that causes the user to feel 'high'.

It is known as a recreational drug, but some users brought to the attention of drug agencies use it on a regular daily basis, and supplement it with the use of aerosols. Hashish oil, several times more powerful than the resin of cannabis itself, can also be abused. There are disagreements about how much damage cannabis taking can produce. When smoking the substance, the smoke is held into the lungs for as long as possible. This, it is suggested, will lead to bronchitis, emphysema, and other respiratory complaints if taken regularly. But there can also be suppression of the body's immune system leading to increased susceptibility to bacterial and viral infections.

I have not found that taking cannabis will lead to other drug use, but certainly taking aerosols and cannabis together appears an increasing problem. Financial consideration may also take the cannabis user on to more addictive drugs. If the user is in a group of people who are involved in drug misuse, whoever is selling cannabis will realise that more money can be made supplying the same people with stronger, more addictive drugs; he will be very persuasive, and he may find one person out of the group susceptible to more addictive drugs.

Sources

Cannabis comes mainly from North Africa, India, South America and Lebanon, but it can also be grown in the United Kingdom. Over the past few years there have been a number of prosecutions for growing cannabis plants. It is very much a recreational, weekend party drug passed from friend to friend.

29

Method of abuse

It is usually smoked in a rolled cigarette with tobacco, but it can also be eaten or brewed in tea.

Effects

Euphoria, altered perception, giggling and laughing, and relaxation are the main effects, but delusions and hallucinations can occur in certain cases.

NUTMEG

Nutmeg is a familiar kitchen spice; that it can be a powerful psychoactive drug will come as a surprise to many people.

Nutmeg is actually the dried seed-kernel of *Myristica fragrans*, an East Indian evergreen tree of the Myristicaceae family. To achieve mild effects, generally 10g (or one-third of an ounce) is sufficient; if this is increased more profound effects can be produced, but the risk of self-poisoning increases (Wells, 1973).

Sources

It can be bought in any good shop; what could be more innocent than asking for nutmeg?

Method of abuse

It is generally taken with fruit juice, but because the solution is emetic it can be difficult to stop vomiting the substance back up again. Once actually swallowed, the user generally has to control the waves of nausea which follow. In this country I have never come across a nutmeg abuser; however, while studying in the United States, I interviewed some older 'drug users' who were completely familiar with nutmeg and used it whenever their own supply of drugs was not forthcoming.

Effects

Euphoria, impairment in concentration, cold, clamminess, dry mouth, and disturbing perceptual effects accompany a 'bad trip'.

PERFUMES

These are usually abused in two particular situations:

(a) When a chronic abuser cannot obtain any other substance he will resort to perfume.
(b) Schoolchildren will take their own or their parents' perfume, put it in a handkerchief and sniff it at school.

It is very rarely abused outside these two situations.

Sources

It is easily bought or taken from the home.

Method of abuse

Abuse is fairly uncommon, and usually only young abusers take perfume, which is generally soaked into a handkerchief and put up to the nose and mouth, or put onto a jacket sleeve and sniffed.

Effects

These are mostly unpleasant: nausea, vomiting, dizziness, sore throat. Some users have reported feeling 'high and good', and a hangover effect. Any excitable behaviour is probably due to the age of the user.

31

PETROL

This is more popular than is generally thought, and used mostly when other solvents are unobtainable, although I have seen several young people in the past few years who only abused petrol.

It is particularly popular in parts of Australia. In the United Kingdom, unfortunately, it contains a high content of lead.

Sources

Petrol is generally stolen.

Method of abuse

It is usually soaked in a handkerchief and sniffed.

Effects

These include headaches, sore throat, nausea, vomiting, dizziness, 'high feeling'. The lead in petrol can cause potential medical problems. There is a hangover effect.

NAIL POLISH REMOVER

Mostly young (12–16 years) females use this substance; it contains acetone and is not very popular.

Sources

It can be obtained from both shop or home.

Method of abuse

It is usually sniffed direct or put on to a handkerchief, or jacket sleeve.

Effects

These include nausea, excitable behaviour (again, possibly due to the age of the abuser), occasional reports of dizziness and headaches.

FIRE EXTINGUISHERS

A small percentage of abusers use this method of obtaining a 'high'. They may find difficulty in obtaining a fire extinguisher (although one of my patients, an enterprising young man in the east end of Glasgow, claimed that over a period of years he used to steal fire extinguishers from a railway yard and sell them to his friends).

Sources

They are usually stolen.

Method of abuse

The contents are generally emptied into cans and bags, then sniffed.

Effects

These include nausea, headaches, dizziness, ataxia, 'high feeling', unconsciousness.

DRY CLEANING AGENTS, PLASTER REMOVER, CORRECTING FLUID THINNER

All three substances are widely abused — see also Appendix B, Table B.4 and Anderson, MacNair and Ramsey (1985).

Sources

They can all be obtained from both shop or home.

Method of abuse

All these are usually taken in a small room (bathroom), one or two cans being released into the air and then breathed in. Other abusers soak a handkerchief or their jacket sleeve with the solvent.

Effects

There is a high feeling, usually accompanied by nausea, with complaints of sore throats and stomach. Dizziness and unconsciousness can follow.

AMYL NITRITE

This was discovered in the 19th century, and used for the treatment of angina pectoris; it dilates the blood vessels leading to the heart, thereby alleviating the pain, and causing blood pressure to be lowered.

In the United States, amyl nitrite has a history of association with the gay community and is a proscribed drug. In the United Kingdom it can be bought over the counter, and has a wider appeal.

Sources

It can be bought normally from a shop.

Method of abuse

It is sniffed straight from the bottle, or small amounts can be soaked into handkerchiefs.

Effects

Some users describe a 'rush' (possibly something to do with the proprietary name). A 'high feeling', slight euphoria, dizziness, headaches and nausea can also occur.

PLASTIC CONTAINERS, BIN BAGS, TABLE TENNIS BALLS

Method of abuse

These are generally burnt over an open fire and the fumes inhaled.

Sources

These can be easily obtained anywhere.

Effects

Since the effects are mainly nausea and dizziness, it is a mystery why these substances are abused.

BANANA SKINS

Over the past few years, 'banana skins' as a substance for abuse has cropped up time and time again. Some habitual substance abusers swear by it, claiming that they get a 'high' and can experience visual hallucinations. Scientists together with drug abuse agencies have also looked at banana skins, and appear to think that it is a gigantic joke (Bozetti, Goldsmith and Ungerleider, 1967). To counteract this argument, it has been said that this line of thinking is to reduce interest in what could be an uncontrollable substance.

NOTE

[1] Butane gas, strictly speaking, is not an aerosol — it can be described as a Gas Fuel. See page 116.

4

Reasons for Solvent Abuse

INTRODUCTION

There are numerous reasons why young people abuse solvents, so at this point it would be prudent to discuss the main causes.

PEER PRESSURE

This is possibly the most important factor. It is not difficult to understand the pressure that young people suffer from their peers in this context. These situations generally arise in large housing estates where considerable numbers of young people of varying ages live.

To an extent, solvent abuse can be described as a temporary fashionable craze. If there is a problem with solvents in one particular area a number of youngsters will experiment, the problem escalates until either the relevant authorities are aware of what is happening or a tragedy occurs, then appropriate action is taken. The 'craze' will then move to another area, leaving behind a few chronic sniffers 'caught in the net'. (Young people living in rural areas are not, of course, immune from this problem; it only takes one practiser to set off a similar pattern).

Most young people will congregate in small groups; so if solvents are being abused in the area, it will quickly become popular for a short while. To be part of the group, to appear 'knowledgeable', possibly emulate their peers, some youngsters will readily abuse solvents.

Unfortunately, at this stage there is some experimentation,

with young people trying to outdo one another, using different solvents, mixing substances (aerosols and tablets, butane gas and cough expectorants etc.; see p. 7).

COST

A large percentage of solvent abusers come from low income backgrounds, and because of the relative cheapness of solvents, with their obvious intoxicating effects, they appear a very inviting proposition to young people who wish to experiment with substance abuse. In addition, the majority of young solvent abusers are still at school, so financial restrictions play an important part in their choice of intoxicating substances.

The home is also an excellent source of freely available substances for abuse: aftershave lotion, perfume, hair lacquer, butane gas, glue, plastic containers. Very often the first indication of a young person abusing solvents will be the disappearance of any of these substances; particularly common is where the butane gas refill can continually needs to be replaced.

One fairly recent noticeable trend is a shift from drug (illegal) use to aerosol or butane gas abuse. Various factors can account for this:

(a) Financial, as the abuser may be unable to raise money for his drugs, and is either unwilling or unable to obtain money illegally for his habit.

(b) There may possibly be a police clamp-down on drug selling (generally amphetamines, cannabis and occasionally heroin), therefore drugs will be unavailable; but aerosols etc. can easily be obtained and abused.

(c) Drugs and solvents, or solvents themselves, can be mixed together.

AVAILABILITY

As mentioned above, the home is an excellent source of available substances, as indeed are building sites (glues), railway yards (fire extinguishers), garden sheds/allotment sheds (paints/thinners).

It is not unusual for some 'enterprising' youngsters to steal large quantities of glue from building sites and then sell small quantities to their peers; every few months this type of behaviour is discovered.

Buying glue/gas, etc. from shopkeepers is an all too common source; in spite of publicity to retailers, there are always a few rogue shopkeepers with an eye to profits, who will willingly sell young people glue, etc.

In the past few years there has been much publicity about the dangers of glue. Certainly, many retailers have adopted a responsible attitude and will not sell glue to young people, but they cannot be aware of all substances that can be abused. (The British Adhesives Manufacturers Association and the British Aerosol Manufacturers Association are very active in this area of prevention, but what could be more innocent than a young person buying a bottle of perfume for his mother!)

In spite of any strategies aimed at stopping the sale of substances to young people, the latter always find opportunities to obtain them.

CULTURAL/SOCIAL

If we look at how substances are abused on a global scale, it will give us insight into how culture can play a part in determining the misuse of substances.

For example, for many years amyl nitrite has had a history of recreational usage in America (particularly among the gay communities in Los Angeles and San Francisco); coca leaves (from which cocaine is extracted) have been chewed by local natives for centuries. In Australia, petrol sniffing is not unknown in the outback; magic mushrooms are popular in Siberia, cannabis in India, paint sprays and lacquers in the poorer, deprived areas of New York. In the United Kingdom alcohol would appear to be the main 'cultural substance'.

It can also be argued that solvent abuse can be influenced by some environmental conditions:

(a) broken home;
(b) below average intelligence (both parents and child);
(c) large family;
(d) low income;
(e) one or both parents with a dependency on alcohol/drugs.

CATEGORIES OF ABUSERS

Attempts have been made in the past to differentiate between categories of abusers, and this is helpful both in deciding a treatment strategy, and dealing with the family situation. For simplicity's sake the four main types of abusers are: experimenter, recreational, chronic and group.

The experimenter

This person will join a group of young people abusing solvents, generally due to peer pressure, or to find out what substance abuse is all about! The vast majority of youngsters in this category only attempt this a few times. They either get no effect, become sick, or — what happens most commonly — they get caught by their parents/teachers, etc. At this stage parents do tend to become overanxious, and fears about their offspring becoming 'addicted' cause them to seek help immediately. The parents themselves find that they are in the midst of bewildering circumstances, feeling guilt, anger, and a sense of failure; this is where intervention by an understanding worker can alleviate the family's understandable fears, and enlist their help and support in treating the youngster. If the worker feels that there are problems in the family that may have assisted in the youngster's substance-taking, this may be the time to enquire into the family's background details. But if initial enquiries are not made, parents can sometimes be more on their guard and will not readily give out personal information. However, the majority of families, after careful and sensitive questioning, will respond with details of their family and any problems they may be experiencing.

Some young people have very low self-esteem, and this coupled with the usual problems associated with adolescence, such as anxiety or depression, can make them prime candidates for abusing solvents, some seeing it as a form of escapism, others as a chance to belong to a group and feel a sense of belonging.

At this juncture it should be noted that although the majority of solvent abusers do give up the habit, not all do, so it should not be viewed as just part of growing up; unfortunately, a small percentage do become chronic abusers.

The recreational user

Just as older teenagers and adults go to the pub for a drink, some young people will justify their usage of solvents with this argument. This type of user will either use solvents on a solitary basis, or as a group activity. The obvious dangers of taking any substance while unaccompanied has potential danger hazards (see Chapter 6), somewhat lessened in group activity.

The recreational user can present several problems if he is to receive appropriate treatment. Unlike the experimenter, this abuser reasons in his own mind that he is fully justified in taking substances: 'They [mother or father] drink, what is the difference with me taking gas, glue, etc.?'

Treatment for this type of user can be extensive (see Chapter 9). A very thin line divides the recreational from the more chronic user, in fact they overlap considerably, particularly when abuse becomes more frequent and is carried on for a long period.

The usual times for taking solvents are generally at the weekend and in the long summer holidays. After school is also a popular time, and when usage is increased, instead of waiting until school is over lunchtime becomes the norm, then the more adventurous will start taking it actually in the classroom.

When this stage is reached, two possible effects can happen:

(1) Some other members of the class will want to join in, thus increasing the problem; somebody always appears to get caught, either at home or at school.

(2) Some of the users' peers are extremely intolerant about abuse, often calling them 'glue head' or 'gas head'. At this point a number of users are found out, and with proper handling of the situation by the teaching profession, and assistance from the appropriate agency, the problem should be resolved.

This stage leaves in its wake some young people who will move on to the more chronic stage.

Gas parties

Gas parties are a fairly new phenomenon, and can be held outside or indoors. The rationale for this type of behaviour is generally example. Youngsters enjoy social occasions; they see their elders having parties with alcohol, so they also have parties, usually with butane gas. There is a mixing of the sexes, loud music, and gas sharing. One of the highlights of the evening appears to be competitions with plastic bags and gas, the winner being the one foolhardy enough to empty a can of gas into a plastic bag covering his whole head and keeping the bag there for the longest time.

On first hearing about this form of abuse, one is filled with horror at the possible fatal consequences. Most, if not all these youngsters will be under the effect of gas, so who is capable, or coherent, enough to help anyone in trouble?

To eradicate this behaviour, particularly if it is unknown to anyone in authority, can be difficult.

The chronic abuser

This group of abuser can be subdivided into two clearly defined types: A and B.

Type A

These are extremely habitual abusers, who can in the main be under the influence of (generally) butane gas for most of the day. They will sniff before school, during school, and certainly all weekend.

If such a chronic abuser is referred for help, great care must be taken to obtain a clearly detailed history of both the youngster and the family background. The chronic abuser is perhaps the most difficult to treat. He may be part of a group, but more often than not he is a solitary sniffer. There may be an exacerbation of pre-existing conditions, for example, major family problems including break-up of family, few friends, or heavy drink abuse in the home.

The helper in this case must be prepared to offer the young person a close, relaxing relationship, and strong encouragement maintained if he relapses, as he invariably will. Young people will sometimes test their helper by deliberately going back onto their solvent to see what type of reaction they can elicit from them ('see, you cannot help me' or 'look I have taken gas again — you do not want to see me').

With this type of abuser, all agencies should be in close contact with each other — school, social worker, treatment personnel, general practitioners and the family. Chronic abusers often have an additional history of petty crime, ranging from theft, housebreaking, car theft, to breach of the peace. Courts are willing to send abusers to treatment centres. (In Scotland, there are children's panels for the under-16s, which can also refer young people to treatment centres.)

One frequent problem that occurs when the youngster attends the treatment centre is that he feels he has been forced to attend, which does not equate with offering a caring service. However, with cooperation from the court and the relevant social (or probation) worker an excellent system can be worked out. It should be explained to the young person that if he is required to attend twice weekly, he really has no option but to attend. However, the helper can offer a 'carrot': if the youngster attends the treatment centre regularly, a request will be made to the appropriate authority for his attendances to be cut. This has a twofold effect:

(a) It allows him to control his own future in a limited way, and gives him some expectation of the future.
(b) He will, hopefully, not see the treatment centre as another prison, and be able to develop a therapeutic relationship with the staff.

It should be borne in mind that very often the chronic abuser has no motivation to abstain from solvents; therefore, any worker who assists youngsters in giving up their habit, must realise that it can be a very frustrating and longterm situation, but with careful treatment planning and helpful encouragement, progress can be made.

Type B

This type of chronic abuser (formerly glue, now mainly butane gas) is the opposite extreme from type A and accounts for some 25 per cent. The astonishing reason for this division in chronic abusers is that after several years of taking solvents, he will voluntarily seek help, generally approaching the agency he has felt most comfortable with. The reasons given for this dramatic about-turn in behaviour are varied, but can be summarised:

(1) He has met a girl and wishes to settle down; very often the partner can be a great source of help and support to the

user, but the worker can also help to keep the user from defaulting.

(2) He is fed up with taking solvents, after several years of obtaining substances through theft, spending all his own money, hiding while taking them.

(3) He is losing friends, the only ones left being generally longterm sniffers as well, so he feels ostracised by non-users.

(4) He is continually getting into trouble with authority — police, school, courts, social workers.

(5) He has family problems — causing upsets at home, perhaps affecting younger members of the family.

(6) He has, perhaps, seen one of his sniffing friends die through solvents. Health problems associated with solvents do not appear to have any effect upon him, although he is more than aware of what they are, but the actual death of somebody relatively close to him through solvents can have a salutary effect.

(7) He has begun to feel a sense of responsibility — very important when treating a user who feels like this. Continual encouragement makes him feel that he is right, and extends the responsibility factor not just to himself, but to others.

In conclusion, this type of chronic abuser who has voluntarily sought help needs encouragement to build a relationship with a counsellor (the user will have several years of solvent-taking behind him, and will need continued support) and reassurance that he has done the right thing in coming forward for help.

Group abuse

Group abuse is by far the most common way of taking solvents, the size of group varying from three or four members up to 12 or 14, of both sexes.

The group meets at regular intervals, which are dependent on the supply of solvents. If it is a very 'sharing' group, whoever has a supply of the solvent will share it with the group.

Fortunately this group activity does not have a long life because of the amount of members; some of them get found out, the group starts to disintegrate and occasionally one or two members may go on to be chronic sniffers.

43

Individual abuse by its very nature, apart from the obvious dangers, can prove to be the most difficult to treat. As we saw above, as it is usually the chronic sniffer who indulges on his own, the reasons for his abuse are more complicated than those of the group sniffer.

When interviewing young people about their solvent problem, alarm bells always ring when they eventually mention that they are in the habit of abusing solvents on their own. It should be stressed that when a youngster is being interviewed by his helper, his 'sniffing history' must be known as soon as possible. This can be difficult on occasions, as gaining trust takes time.

5

Legal Issues in Solvent Abuse

INTRODUCTION

Legal issues can be a very complex subject. To enable the reader to comprehend what laws govern solvent abuse we must look at both Scotland and England for an overall view of these aspects.

SOLVENT ABUSE ACTS

Solvent Abuse (Scotland) Act 1983

This Act does not make it a criminal offence to sell or abuse solvents, but it does create an additional ground on which a child may be referred to the Reporter to the Children's Panel as being in need of compulsory measures of care (Social Work (Scotland) Act 1968, Section 32 (2) (gg)). A child who is found by police officers under the influence of solvents or in the act of inhaling the vapour of a volatile substance may now be referred.

Children's panels

The following information is published by the Scottish Information Office, Fact Sheet 7. In Scotland, children who commit offences, other than certain specified offences, or who may be in need of care and protection go before a 'children's hearing' rather than a juvenile court. A child may be brought before a children's hearing for: being beyond the control of parents; falling into bad association; being exposed to moral danger; or being caused unnecessary suffering or serious impairment to health or development through lack of parental care.

Children under 16 continue to appear in court where serious offences such as murder or assault to the danger of life are in question, or where they are jointly accused with adults, or are involved in offences where driving disqualification is possible, or forfeiture of weapons necessary.

The Reporter

The central figure in the system is a local authority official called the Reporter. Information is passed to him by police, social workers, teachers and others who may include general members of the public, who know of children in difficulty. Whatever the reasons for the referral, the Reporter may decide between three courses after making the necessary investigations:

(a) no formal action, on the grounds that compulsory measures of care are not considered necessary;
(b) voluntary referral to the social work department for advice, guidance and assistance;
(c) referral to children's hearing on the grounds that, in his view, the child is in need of compulsory measures of care.

Members of the children's panel are recruited from people in a wide range of occupations, neighbourhood, age and income groups. There is an approximate balance between men and women and a wide range of ages (20–60). All have experience of and interest in children, and the ability to communicate with them and their families. Members are carefully prepared for their task through initial training programmes and have continuing opportunities during their period of service to develop their knowledge and skills and attend inservice training courses.

Case history

Over the past few years various newspapers have reported on the prosecution of shopkeepers who have sold solvents to young people for the purpose of intoxication. One such case occurred in Glasgow where a shop was visited on several occasions by police officers who attempted to reason with the occupiers about the moral obligation they had regarding the sale of solvents to children (persons under 16 years of age).

After many consultations, it became apparent that no heed was being taken and the sales continued. The police then prepared a case obtaining statements from all known children who had obtained solvents at these premises. The investigations revealed that a number of children were being supplied with plastic bags (food bags) on purchasing the solvents and also that a number of the children were exchanging stolen goods for solvents.

In legal terminology, both persons were charged with culpably, wilfully and recklessly supplying quantities of solvents (in particular Evostik glue), in or together with containers such as tins, tubes, crisp packets and plastic bags, for the purpose of inhalation of the vapours of said solvents from within said containers for the said purpose, and that the inhalation by said children of the vapours of said solvents was or could be injurious to the health of said children and to the danger of their health and lives.

Both persons were further charged with reset (handling stolen goods). The main charge was in fact an extension of the existing Common Law of Scotland, which is non-statutory and is flexible to allow change to present-day trends, etc.

The relevancy of this charge was contested by defending counsel and the first appeal on the relevancy of the charge was heard before one judge who upheld that the charge was relevant. A further appeal was made in front of three judges who also upheld that the charge was relevant. Both accused appeared for trial at Glasgow High Court. (Mr. A. Vauet of the Crown Office, Edinburgh).

Intoxicating Substances Supply Act 1985

It is an offence for a person to supply or offer to supply a substance other than a controlled drug:

(a) to a person under the age of 18 whom he knows or has reasonable cause to believe, to be under that age; or,

47

   (b) to a person:
      (i) who is acting on behalf of a person under that age; and,
      (ii) whom he knows, or has had reasonable cause to believe, to be so acting,

if he knows or has reasonable cause to believe that the substance is, or its fumes are likely to be inhaled by the person under the age of 18 for the purpose of causing intoxication.

A person guilty of an offence under this section shall be liable on summary conviction to imprisonment for a term not exceeding six months or to a fine not exceeding £2000 or both.

Magic mushrooms

The laws pertaining to magic mushrooms can be very complex. For example, there is no law pertinent to the harvesting, preparation or use of fly agaric (*Amanita muscaria*). The following legal issues are from the ISDD drug abuse briefing.

With mushrooms containing psilocin and psilocybin, these substances are controlled under class A of the Misuse of Drugs Act, and under regulations that prohibit their medical use; their possession, production or supply, or the act of allowing premises to be used for their production or supply, are offences, unless in accordance with a Home Office licence issued for research or other special purposes.

The drugs are not illegal while they are in the mushrooms, until the latter are boiled or crushed to make what is known as a 'preparation or other product' containing psilocin or psilocybin. Such a preparation is a controlled substance, subject to the same restrictions and penalties as the drug it contains.

Cannabis

Cannabis in its various forms is controlled by the Misuse of Drugs Act, under regulations that prohibit both its medical and non-medical use. Therefore, it is illegal to cultivate, produce, supply or possess the drug, except in accordance with a Home Office licence issued for research or other special purposes. It is also an offence to allow premises to be used for producing (including cultivating), supplying or smoking cannabis.

This last type of offence, allowing the use of a drug, applies only to the smoking of cannabis or opium. Herbal cannabis (everything except seeds and stalks), cannabis resin, and (since it is prepared from the resin) cannabis oil are in class B of the Act. Active chemical ingredients (cannabinoids) that have been separated from plant (generally unlikely) count as class A drugs.

Amyl nitrite

At present there is no law governing this substance, and it can be bought quite openly.

Lysergic acid diethylamide (LSD)

LSD is controlled under class A of the Misuse of Drugs Act under a set of regulations which prohibit both medical and non-medical use. This means it can only be supplied or possessed for research or other special purposes by persons licensed by the Home Secretary. Other than in these limited instances, production, supply and possession of LSD and other hallucinogens is an offence under the Act. It is also an offence to allow premises to be used for the production or supply of these drugs.

ACTS REGULATING AVAILABILITY OF DRUGS

There are two main statutes regulating the availability of drugs in the United Kingdom.

The Medicines Act 1968

This Act governs the manufacture and supply of medicinal products (mainly drugs) of all kinds, and its enforcement rarely affects the general public. It divides drugs into three categories: the most restricted (prescription only) can only be sold (or supplied in 'circumstances corresponding to retail sale') by a pharmacist working from a registered pharmacy, and then only if the drugs have been prescribed by a doctor; the least restricted (general sales list) can be sold without a prescription by any shop, not just a pharmacy, but even here, certain advertising, labelling and production restrictions apply; all the remaining products (pharmacy medicines) can be sold without a prescription, but only by a pharmacist.

This is the only law controlling tranquillisers, all of which are obtainable on prescription only.

49

Misuse of Drugs Act 1971

This Act is intended to prevent the non-medical use of certain drugs. For this reason it controls not just medicinal drugs (which are also in the Medicines Act), but also drugs with no current medical uses. Offences under this Act overwhelmingly involve the general public, and even when the same drug and a similar offence is involved, penalties are far tougher. Drugs subject to this Act are known as 'controlled' drugs.

The law defines a series of offences, including unlawful supply, intent to supply, import or export (all these are collectively known as 'trafficking' offences), and unlawful production. Unlike the Medicines Act, 'supply' here encompasses giving drugs away, as well as selling them. But the main difference is that the Misuse of Drugs Act also prohibits unlawful possession. To enforce this law the police have the special power to stop, detain and search people on 'reasonable suspicion' that they are in possession of a controlled drug.

Penalties

Maximum sentences differ according to the nature of the offence: less for possession; more for trafficking, production, or for allowing premises to be used for producing or supplying drugs.

They also vary according to how harmful the drug is thought to be. Class A drugs have the highest penalties (7 years plus unlimited fine for possession; 14 years plus fine for production or trafficking), and include the more potent of the opioid painkillers, hallucinogens and cocaine.

Class B drugs have lower maximum penalties for possession (5 years plus fine) and include cannabis, less potent opioids, strong synthetic stimulants, and sedatives.

Class C drugs have the lowest penalties (2 years plus fine for possession; 5 years plus fine for trafficking), and include some less potent stimulants, and dextropropoxyphene, a mild opioid analgesic (with paracetamol known as Distalgesic).

Any Class B drug prepared for injection counts as Class A.

Less serious offences are usually dealt with by magistrates' courts, where sentences cannot exceed 6 months, plus £1000 fine or 3 months plus fine for less serious offences.

Over 80 per cent of all drug offenders are convicted of unlawful possession. Although maximum penalties are severe, just one in six offenders receives a custodial sentence (even fewer actually go to prison) and nearly three-quarters of fines are £50 or less.

Regulations

Regulations made under the Act divide the controlled drugs up in a different way to take account of the needs of medical practice. In effect, they define exceptions to the general prohibitions on possession, supply, etc. The most restricted drugs can only be supplied or possessed for research or other special purposes by people licensed by the Home Office; these drugs (for example, cannabis and LSD) are not available for normal medical uses and cannot be prescribed by doctors who do not have a licence.

All the other drugs *are* available for normal medical uses (for example, strong analgesics like morphine, stimulants like amphetamines or cocaine, and some sedatives), but most are prescription-only, so they can only be obtained if they have been prescribed by a doctor and supplied by a pharmacy. Some very dilute, non-injectable preparations of controlled drugs (for example, some cough mixtures and antidiarrhoeal mixtures containing opiates), because misuse is very unlikely, can be bought over the counter without a prescription, but only from a pharmacy.

Additional regulations effectively restrict the ability to prescribe heroin, dipipanone (Diconal) and cocaine for the treatment of addiction to a few specially licensed doctors.

OTHER ACTS RELATING TO DRUGS

Customs and Excise Act

Together with the Misuse of Drugs Act, the Customs and Excise Act penalises unauthorised import or export of controlled drugs. The maximum penalties are the same as for other trafficking offences, except that fines in magistrates' courts can reach three times the value of the drugs seized.

Road Traffic Act

Road traffic legislation makes it an offence to be in charge of a motor vehicle while 'unfit to drive through drink or drugs', the word 'drugs' here including both prescribed drugs and solvents.

6

Physical and Psychological Effects of Solvent Abuse

INTRODUCTION

Physical and psychological dependence on solvents were briefly dealt with in Chapter 2 and are now covered here in considerably more detail.

PHYSICAL

Medical problems

There is often some overreaction in discussion of medical problems associated with solvent abuse. Certainly there are medical complications with the misuse of solvents, but they should be taken in the correct perspective with the overall number of young people who take solvents.

If the abuser is a solitary glue-sniffer, for example, particularly if he puts a plastic bag over his head to enhance the effect, asphyxiation can occur.

Polysolvent use, or mixing drugs/alcohol with solvents, can also be extremely risky due to the potentiating effects of these substances. Asquith and Didcott (1983) mention that the 'chronic' long-term abuser,

> whether of toluene or other solvents, runs a much greater risk of toxic injury, particularly to the central nervous system, than the short-term or casual user. This may manifest in physical effects such as trembling (ataxia) and the loss of fine motor control, and psychological impairment

such as short-term memory loss and intellectual impairment. These effects may be reversible unless the damage is severe. The chronic solvent abuser tends to become dependent upon solvents and may become a withdrawn, isolated and solitary individual with no motivation to stop sniffing. However, the number of extreme chronic abusers is very small indeed in proportion to the total number of persons who abuse solvents at one time or another.

Other reports have been published showing some dangers of solvent abuse. The following references deal with medical problems encountered in solvent misuse: Anderson, MacNair and Ramsey, 1985; Anderson *et al.*, 1986; Akerman, 1982; Baselt and Cravey, 1968; Cohen, 1973; Clark and Tinston, 1982; Francis *et al.*, 1982; Fagan and Forest, 1977; Flowers and Horan, 1972; Garriott and Petty, 1980; Haq and Hameli, 1980; Herd, Lipsky and Martin, 1974; Kringsholm, 1980; Mee and Wright, 1980; Poklis, 1976; Roberts, 1982; Reinhardt *et al.*, 1971; Taylor and Harris, 1970; Weissberg and Green, 1979.

Fatalities

There have been occasional reports of young solvent abusers dying after a session. It would appear that violent exertion also increases the possibility of death.

Bass (1970) suggests that some volatile substances sensitise the heart to the effects of sympathetic stimulation which may occur during sudden exercise and other activities. Other studies have shown that in human subjects exposed to chronic industrial levels inhalation may cause impairment of psycho-motor performance and a slowing of reaction time (Waldron, 1981). (See Chapter 9: treatment of solvent users with the aid of a computer (p. 88), shows visually the slow reaction time some users demonstrate; after ceasing abuse, reaction time improves.) There is an excellent comprehensive report on solvent misuse by the National Institute on Drug Abuse (1978), which include the facts that lead in petrol may produce brain damage; dry-cleaning fluids containing trichlorethylene and trichlorethane may give rise to renal or hepatic damage; and butane gas is cardiotoxic.

Obviously, there remains a great deal of research needed to study the problems of solvent misuse.

One of the greatest difficulties is the follow-up of the solvent abuser, assuming that he has come to the attention of some form of treatment centre or other authority. Another difficulty is that no-one is sure exactly how many young people have actually abused solvents, albeit only experimentally, and there must be a number of chronic sniffers who have never come to the attention of any authority.

Wiseman and Banim (1987) have reported on a 15-year-old boy with a two-year history of intermittent solvent abuse who underwent cardiac transplantation. Cunningham *et al.* (1987) cited a case of myocardial infarction and primary ventricular fibrillation after glue-sniffing.

King, Smialek and Troutman (1985) have reported on sudden death in adolescents resulting from the inhalation of typewriter correction fluid.

Indirect dangers

Because of the fear of being caught, some solvent users, both group and solitary sniffers, tend to sniff in deserted areas such as old disused buildings, canal banks, and railway sidings, where dangers include:

(a) asphyxiation: if group-based, other members may take fright and run off;
(b) choking, vomiting: once again members can take fright and run off;
(c) violence: under the influence of the solvents, the user may become violent, or paranoid, possibly due to 'seeing things';
(d) falling into a river or canal;
(e) falling out of buildings or high flats.
(f) if near a road, the added danger of wandering in front of traffic. Two years ago a Glasgow newspaper carried the story of two five-year-old boys wandering in front of cars on a busy main road, both under the influence of solvents.

Physical symptoms

The most common physical symptoms associated with solvent abuse are:

(a) nausea
(b) vomiting, diarrhoea
(c) headaches
(d) stomach pains
(e) backache
(f) ataxia
(g) sweating
(h) drowsiness
(i) unconsciousness.

Emergency treatment

It is important not to panic for two reasons:

(a) You may not take the most effective action.
(b) If you alarm the sniffer, flight or violent reaction may cause arrhythmias (leading to heart failure).

Immediate action

(a) Remove the solvent or other substance.
(b) Remove anything which may cause asphyxia.
(c) Provide copious fresh air.
(d) If in doubt call an ambulance.
(e) *Definitely* call an ambulance if he/she is unconscious.

Interim action

(a) Lay the person down on his/her stomach, with his/her head to one side to prevent inhaling vomit; ensure that the tongue is not obstructing the airway.
(b) Loosen all tight clothing.
(c) Do not give anything to eat or drink.
(d) Remain with the person until conscious.

This advice is given by Re-Solv, and is contained in their information pack.

If the person is not breathing, give the 'kiss of life':

(a) The person should be turned on his/her back, if possible.
(b) Check that there is a clear airway — no obstruction by vomit or dentures.

(c) Bend the head well back and try and support with folded clothing.

(d) Pull the jaw forward.

(e) Keep the nostrils closed with thumb and forefinger.

(f) Cover his/her mouth with your own.

(g) Blow in until the chest fills.

(h) Observe it deflate.

(i) Repeat (f), (g) and (h) every time the chest deflates, but not more than 20 times per minute.

This procedure should be carried out until professional help is available.

Afterwards

Encourage the person to talk about reasons for sniffing, particular worries and problems.

PSYCHOLOGICAL

The chronic user of solvents (fortunately in a minority) generally has both emotional and psychological problems. The solvent taking is a symptom rather than a cause of his main problems. He may have started sniffing in an attempt to gain acceptance with his peers (usually an isolated individual); or he has parents who have an alcohol/drug problem and he gravitates towards solvents (cheap and easier to obtain); or he may be using solvents all his waking hours, as a crutch. Whatever the reasons are, this type of solvent abuser will require intensive treatment. In other types of solvent abusers, the habit is usually transitory, and there are few psychological problems requiring help.

An interesting paper on the cognitive functioning of solvent abusers was published by Dr Z. Mahmood a few years ago (Mahmood, 1983), and pursues an interesting aspect of solvent abuse.

Cognitive dysfunction and solvent abuse have been found to be closely related, but the precise nature of this relationship remains obscure. Twenty-eight non-active solvent abusing and twenty non-abusing adolescents were given tests of non-

57

verbal and verbal intelligence, arithmetic, literacy level and learning ability. In addition, they were asked three standard questions about their sparetime activities and future orientation. The results clearly indicated that, while the two groups did not significantly differ in potential abilities, the abusers performed significantly poorly on all but one test of acquired abilities. The abusers were found to be less future-orientated, were 'bored' with themselves.

The main psychological problems that arise are:

(a) lowering of inhibitions
(b) emotional instability
(c) aggressive behaviour
(d) mild psychotic behaviour
(e) hallucinations.

Part II

Nursing and Community Aspects

7

Nursing Intervention
in a Clinical Setting

INTRODUCTION

Due to her unique position of being present in a number of
facilities where solvent abusers may be referred for medical
assistance, the nurse should not only have knowledge about the
subject, she should be aware of what facilities are available, and
how to contact them. There must obviously be extensive
advertising by the agencies concerned, and hopefully, they will
realise how important a contact the nurse can be, and the
number of areas with which she will be connected: accident and
emergency units, general practitioners' surgeries and health
centres.

ACCIDENT AND EMERGENCY DEPARTMENTS

Over the past few years there has been a steady increase in the
number of young people attending accident and emergency
departments with medical problems associated with solvent
abuse. They arrive in a variety of conditions, from those
accompanied by worried parents not knowing whom to turn to
for help when they discover their child abusing solvents, to
emergency admissions where the solvent abuser has been found
choking or unconscious.

Educating the nurse about solvent abuse should start in her
training, particularly if she is taking the accident and emergency
course. She must know:

(a) what solvents are;

(b) types of solvents;
(c) signs and symptoms of a youngster who has taken solvents;
(d) how to handle an abuser who is intoxicated, and showing signs of aggression, fear, hallucinations;
(e) relevant referral agencies;
(f) relevant support groups for parents.

Because of the workload at accident and emergency departments, the nurse will not usually have the time or opportunity to contribute towards the actual treatment of the solvent abuse problem, only provide emergency care; therefore, it is important that adequate liaison exists between agencies to give accident and emergency departments information regarding facilities available for solvent abusers. These agencies may provide study days, seminars and information which can be made available for accident and emergency staff. Any interested nurses should be encouraged to attend these treatment centres and see what is available.

GENERAL PRACTITIONERS' SURGERIES AND HEALTH CENTRES

Often when parents are confronted with solvent abuse in the family, the general practitioner is the person they turn to for help. However, the general practitioner is generally very busy, and after physical examination his time will be limited as regards counselling and follow-up care.

Here the practice nurse, or other nurses in the health centre, health visitors, or community psychiatric nurses either may be able to offer counselling and support, or will be able to refer the solvent abuser to a helping agency.

THE IMPORTANCE OF WORKING TOGETHER

Experience has shown that where there is an exchange of information and ideas, performance and ability can increase, as does knowledge, particularly in dealing with the ever-changing problem of solvent abuse. Different solvents are abused periodically; in some areas gas can be the main substance, then very suddenly this can change to pain relief sprays. If only to discover what 'new' substances are being abused, all relevant agencies and workers must have close liaison. Different workers and treatment centres all have separate methods of treatment. They may be similar, but there will be differences; for example, some may offer specialised forms of treatment, such as hypnosis, which would be unavailable elsewhere.

Hopefully then, in any given region there should be a variety of treatments, workers, and treatment clinics or centres available for treating solvent abusers.

How does the worker discover where these places are, and what they have to offer? One method that has been successful is a Register of Addiction Services in Strathclyde issued in booklet form by the Social Work Department, and compiled by the Information and Resource Unit on Addiction in Glasgow. Part of this Unit's remit is to provide up-to-date information regarding services available to deal with problems of addiction in Strathclyde. With the publishing of this Register, there is now a readymade source of material available giving:

(a) contact person, address, telephone number;
(b) opening hours of services;
(c) governing body;
(d) services provided;
(e) method of referral;
(f) cost to client.

Other organisations have similar systems in operation, detailing available agencies for the treatment of alcohol, solvents and drugs: Re-Solv, the Society for the Prevention of Solvent and Volatile Substance Abuse, and SCODA, Standing Conference on Drug Abuse, for example.

This method of information sharing is of great benefit to all, as by providing useful addresses and methods of referral, the worker is able to refer his clients to the most appropriate agency.

Another important step in assisting cooperation between agencies would be the establishment of a forum which would consist of interested workers in the related field, hopefully both professional and voluntary. The aims of the forum could be:

(a) Meet regularly.

63

(b) Draw up a committee.

(c) Advertise the forum's aims.

(d) Exchange ideas and experiences.

(e) Set up various training schemes for workers to extend their knowledge.

(f) Offer assistance to groups wishing to establish a treatment centre, or to self-help groups.

(g) Issue three-monthly newsletters containing the latest developments, relevant information and addresses.

(h) Organise publicity, not only for services available, but also for a general understanding about solvent abuse and how the public can help. This can be done by contacting newspapers, offering interested groups talks on the problems of solvent abuse, and organising seminars.

GIVING SUPPORT TO THE FAMILY

At this time, the family members may also require support and guidance, in what can be a traumatic happening for them.

Before the solvent abuser was brought to the attention of a helping agency, the family may have had several harrowing experiences: the child not coming home, showing erratic behaviour, causing worry and fear; the child being involved with the police. The parents may feel guilt, and there may be other problems in the family: financial worries, unemployment, single-parent family, alcohol/drug problems. The solvent abuser may be reacting against problems in the home, or he may be under peer pressure to take solvents outside the home.

The importance of involving the family cannot be emphasised enough. Even if the worker is certain that the client does not have other problems, including family ones, the family background should be investigated. It might well turn out that all the family requires is support for themselves while the problem is being treated. However, in a number of cases, family work may be required. The majority of parents, when confronted with their child's misuse of solvents, react with horror that their child may be suffering from various illnesses and may even die.

Certainly the use of solvents can lead to medical problems, and unfortunately death (see Chapter 6), but to put the problem into its correct perspective, the vast majority of solvent users appear to remain unscathed from their experiences.

This should be explained to the parents, and also the reasons why young people abuse solvents. The most useful analogy that can be used as an example is to explain that when they themselves were younger and at school, smoking cigarettes was considered very risqué, and on a number of occasions peer pressure would be at work. Today it is solvents, not cigarettes.

Of course, as with any experimentation in substances, there will always be a percentage whose use of solvents will start becoming a problem in their lives.

Family work

Establishing a relationship with the family with the intention of working therapeutically requires great skill and well-experienced staff. If the worker involved realises that the youngster is in the category of having quite severe problems at home, family work, whenever possible, should be attempted. It is beyond the remit of this book to guide workers through what can be a traumatic experience for the family being involved in therapy when there are definite family problems. The services of a clinical psychologist or psychiatrist can be obtained.

If, however, the client's worker is of the opinion that both the family's and the client's problems can be discussed with him without too much trauma, it should be attempted, provided the worker has the relevant experience.

Family work is basically exploring relationships in the family and the interaction between members. Some problems, alcoholic or financial, for instance, may be very much in evidence, others may not. Careful questioning of the parents' attitude to the youngster involved, and their attitude to each other, is vital; one parent may be played off against the other. Other members of the family should also be involved. What is their attitude towards their brother or sister being involved with solvents? Are all members treated fairly equally by the parents, or is the solvent abuser being made a scapegoat for the family's problems? Perhaps his low self-esteem comes from the family's attitude towards him, and the more he becomes involved with solvents and the attendant problems, the more the attitude of the family goes against him; so the lower his self-esteem will sink, and he will become even more isolated.

Other agencies that may have been involved with the client or

65

the family should be involved in discussions about the family situation, and what they can offer in any capacity.

Family work, as we have seen, can be an extremely complex and time-consuming activity. It should not be taken lightly, but if the worker is convinced that this is the only course of action to help the client and build up relationships within the family for him, then family work should be undertaken.

8

How Community Services Can Help

INTRODUCTION

It would be unfair to state that the problem of solvent abuse should be tackled by one professional body. Indeed, because of the complexities involved in such cases, there is a clear need for a multidisciplinary approach. However, for various reasons, this is not always available, so what is the individual practitioner's strategy? Also, there has always been a degree of confusion as to who will treat solvent abusers. All workers in the 'caring environment' will come into contact with this type of client, and often it will be the individual's choice if he feels he has the necessary skills for helping/treating the abuse. For the acute stage, the client can be referred to a local clinic, and the worker in conjunction with the clinic can follow up the case and offer support and guidance.

Unfortunately, there is a dearth of clinics available for this method; so the worker may have to plan his own strategy with the client, hopefully with help from more experienced colleagues, and from other professionals who may have a working knowledge of the associated problems.

A number of different professionals work in the community, and the solvent abuser will probably come to the attention of these groups:

(a) social workers;
(b) health visitors;
(c) community education workers;
(d) teachers in schools and further education.

67

SOCIAL WORKERS

In the past few years, social workers have been providing various much-needed services for young solvent abusers.

Some workers conduct their 'treatment' on the street. This basically entails talking to young people in their own environment, and can be done anywhere, from street corner to parks, cafes or any place where young people congregate. This approach does have some success, but obviously the worker has to develop a close rapport between himself and the potential client. Without the support of a clinic or fellow workers, the social worker has an extremely difficult job, probably the most demanding of all strategies aimed at helping solvent abusers.

Once trust has been established, the worker then has the task of attempting to curtail the abuse of solvents. Even if the client does not wish to stop taking them, the worker should still offer potential help, listen to the client talking about personal problems, and the relationship will continue. The worker is not condoning the use of solvents, but very often in these circumstances this type of client will certainly not wish to attend clinics; this method of approach can be said to have some therapeutic value, and the client in this category generally does stop abusing solvents, the main reasons being that he/she:

(a) actually grows out of the habit;
(b) begins a relationship with the opposite (or same) sex;
(c) becomes fed up with numerous brushes with the law;
(d) is encouraged to pursue other interests, sport or hobbies;
(e) develops a therapeutic relationship with the worker and a degree of trust is present — at this stage the user may wish to demonstrate his willingness to stop abusing solvents to gain the worker's respect.

There is always a small percentage who will, in spite of intervention, continue with solvents and may go on to other substances. All the worker can do in this situation is once again to be available if the young person wishes to talk or requires support.

Intermediate treatment centres have been opened, operated by social work departments, offering facilities for young solvent abusers to attend; some treatments are available and leisure activities are encouraged.

Solvent abusers come to the attention of social workers through a variety of other agencies — children's panel (Scotland), juvenile courts, police — who are looking for guidance and help with them.

Residential schools

In residential schools where young people have been admitted for a variety of reasons, solvent-taking may be a main cause, or they may be involved in other activities, and solvent-taking is only peripheral; treatment can be offered to them.

Unfortunately, solvent-taking in some circumstances can increase when youngsters of varying ages and offences are living together; even previous non-users may be tempted into abuse. This type of problem is not just confined to residential schools.

Various strategies are available to the workers in residential schools. One method frequently used is to refer the solvent abuser to a local clinic for the acute stage and, with close cooperation from both areas regarding continuity of care, the youngster can then be supported by the residential staff. If this method is impractical, the residential staff should pursue other strategies. Some staff may feel they have very little experience or knowledge about the subject. This can be rectified; inservice training can be initiated, provided by staff already working in the field of solvent abuse, and residential staff can be seconded to treatment centres. Information can be obtained from advisory services, and courses are available for staff wishing to extend their knowledge in the solvent and drug field.

Overall, the social worker is in a strong position to help the solvent abuser, be it on the street, through residential settings, intermediate treatment centres, or 'drop in' centres.

HEALTH VISITORS

Included in this section, because of their potential proximity to solvent abuse, either directly or for advice, are district nurses, community psychiatric nurses and nurses in health education.

All health workers have experience and knowledge of health education, and are often called upon to give advice and assistance to families who have a child abusing solvents.

With the rising problem of solvent abuse, knowledge of the subject and how it can affect the family is essential for all health personnel. Fortunately, in some training colleges the subject is being taught; for those already working, it would be hoped that through inservice training, seminars, etc., the worker is able to recognise the problem and take the appropriate steps.

The actual physical areas that nurses cover are very large, from general practitioners' surgeries to health centres, and working in the community and in schools, so most nurses would have some contact with a young person involved with solvents.

However, if a treatment centre is unavailable locally for these clients, how will the youngster's problem be solved?

If the nurse has the knowledge and experience to deal with the problem, she may be able to offer regular counselling and support. The client should be medically examined, his/her own general practitioner should be involved, and after discussion with other relevant authorities who may also be involved, a programme can be established.

This method may have some success, as the majority of young solvent abusers require support, a person to whom they can relate, somebody they trust with whom to discuss problems, and help increase their self-esteem.

Obviously there will be exceptions, and other methods may be required. If the solvent-taking is increasing and there are other serious associated problems, some type of residential care may be appropriate, with assistance from a psychologist or, if needed, a psychiatrist.

In schools and colleges, the likelihood of problems with solvents is increased, with large numbers of young people present, peer pressure possibly very strong, and a desire to experiment. Once solvents appear the problem can escalate rapidly. Other young people will wish to try, and within a short period of time, a fair proportion of youngsters will have experimented. Fortunately the 'craze' will die out, usually as quickly as it commenced. However, the nurse attached to a school or college should be able to deal with this problem if it is referred to her. She should be aware of what facilities, clinics or treatment centres, are available for assistance. And, of course, she must be able to deal with medical emergencies, such as choking, vomiting or unconsciousness.

She will be the first step either in treating the client or for referring him/her elsewhere, so either through her early or inservice training, she must gain a knowledge of solvents, and the appropriate steps to take when confronted with this problem.

Nursing organisations are concerned with the abuse of substances, and through the Royal College of Nursing a forum on substance abuse was initiated in 1986. Various articles have been published in the nursing periodicals outlining solvent abuse and the strategies being taken to combat it.

COMMUNITY EDUCATION WORKERS

Community education workers have an important role to play; they are involved at local level with a variety of projects and work with volunteers and part-time youth workers.

The community worker can assist in combating solvent abuse, from a preventive view, through working in youth clubs, using such activities as discussion groups, encouragement of appropriate recreational games, hobbies, social education. Outside speakers can be invited to talk about their work in this area, and the health aspects.

The community worker is in an ideal position to help young solvent abusers; being involved in leisure pursuits for young people, he can offer trust and friendship, and will listen to any problems the young person is having, give advice if required, and help build self-confidence.

If the community worker feels he does not have the relevant expertise to offer such counselling, he can liaise with agencies already involved in treating solvent abusers, or any other source with some experience of counselling them.

Seminars, inservice training, organising local coordinating groups of different workers, and supporting local groups such as self-help and parents' groups, can all be organised by the community education worker to increase general awareness of the problems of solvent abuse and where help and guidance can be given.

TEACHERS IN SCHOOLS AND FURTHER EDUCATION

Given that most young people under the age of 16 will spend a large proportion of their time within the school environment, there is a strong possibility that some form of solvent-taking will go on during school hours. Certainly teachers should be aware of the signs and symptoms of solvent abuse at school:

(a) persistent truancy;
(b) irrational, unusual behaviour;
(c) outbursts of laughter, giggling for no apparent reason;
(d) no interest in the surroundings;
(e) unsteady gait, stumbling, uncoordinated movements;
(f) drowsiness;
(g) perhaps continually sniffing the sleeve (it is not unusual for some youngsters to sniff a substance in class that has been sprayed or poured on to their sleeves).

Drugwise 12–14

This is a Scottish Drug Education Programme for 12–14-year-olds.

One strategy initiated by the Director of the Education Department in Strathclyde was to set up a working party to investigate the suitability of existing teaching materials on drug education. They concluded that there were *no* existing materials which provided teachers with effective resources.

A regional development team was then started up to produce teaching and learning materials for initial piloting within the Strathclyde region, in its first phase, but intended for wider dissemination throughout Scotland in its second phase.

The Scottish Education Department issued a circular to the education authorities encouraging them to become involved in initiatives to combat drug misuse. Funding for the two initiatives by the Strathclyde Education Department was formally announced in the circular:

(a) package of new teaching materials on drug misuse to be used in schools;
(b) provision of a grant to education authorities for inservice training of staff.

The focus of the programme is on the individual. The 'shock horror' approach is minimised and the information supplied is limited, but accurate. This approach has been supported by various research projects in drug education. In addition, experience has shown that it is more likely that young people will be offered drugs for the first time by friends or acquaintances, not by 'evil drug pushers', so the programme endeavours to prepare them for this situation by giving them the appropriate skills:

(a) decision-making;
(b) rejection (saying '*no*');
(c) dissuasion (helping a friend say '*no*').

Time management is also included, where using their free time in a constructive manner should lessen the risk of pupils being offered, accepting or using drugs.

In addition, Strathclyde police have developed a short programme which gives limited but accurate information on drugs:

(a) presentation on types of drugs abused in Scotland with aspects on the law regarding illegal possession and supply of controlled drugs;
(b) case history illustration on the legal process of a person who is suspected of supplying drugs (to involve the pupils as much as possible it is accompanied by a cartoon strip);
(c) question and answer session.

This involves both officer and teacher taking an active part, stimulating and encouraging questions and debate. The decision whether or not to take drugs is influenced by many factors, and attempts have been made to heighten the pupils' awareness of this, where appropriate.

By the end of the programme, which is directed towards pupils who have not previously become involved in drug misuse, they will have had the opportunity to:

(a) extend their knowledge of drugs;
(b) understand why some drugs are socially acceptable and some are not;
(c) understand the legal consequences of using controlled drugs;
(d) be more aware of their own thinking about drugs;

73

(e) consider some of the influences that might affect their responses to an offer of controlled drugs;

(f) consider and practise some of the skills that might be helpful in refusing an offer of drugs;

(g) explore new ways of using their free time.

The work of the programme in developing decision-making, rejection and dissuasion skills is in the context of drug offer situations. For skills to be acquired, it is necessary to practise them, and this is achieved through role-play where appropriate.

Decision-making skills

This highlights the types of information used by an individual in making an informed choice about the use of drugs.

Rejection skills

This includes showing how the circumstances in which an offer of drugs is made can affect how easy or how difficult it is to say 'no'. These circumstances are explored by the pupils, focusing on skills which may help a pupil say 'no' to an offer of drugs.

Dissuasion skills

This focuses on friendship and on ways in which pupils may help a friend say 'no' to an offer of drugs.

Parental involvement

Because of the concern parents may have, and their role, it is advisable for them to be fully informed about any drug education taking place.

Drugwise 12–14 was introduced into Strathclyde schools in January 1987, and is currently being piloted in other Scottish regions. Previously, Drugwise 14–18[1] (the titles are coincidental) had been produced independently in England, partly sponsored by the Scottish Health Education Group, and hopefully it will be used in Scottish schools.

Although aimed at drug misuse, the Drugwise Project also demonstrates an up-to-date strategy which is being used today, and how an efficient programme can be initiated, worked through, and presented to the appropriate audience, for both education and active participation.

Currently available are a number of excellent teaching 'aids' dealing specifically with solvent abuse. A comprehensive list appears in Appendix B.

NOTE

1. Available from Tacade, Trafford Road, Salford M52 XS, price £34.

9

Treatments

INTRODUCTION

A number of treatments are available for helping solvent abusers to cut down, abstain and remain solvent-free. No treatment is intended to be taken in isolation. Care should be taken when deciding what programme is suitable for the individual, and ideally this should be discussed with at least two members of the team.

MIXING AGE GROUPS

Another important aspect is that some clients can be very young (maybe seven or eight years old), so care must be exercised that they do not have the opportunity to mix with some of the older clients (teenagers and some perhaps in their twenties) who may unduly influence them, particularly if the latter are still abusing solvents. This is not to say that mixing age groups is completely undesirable. In certain controlled areas where there is positive interaction between age groups, and no possibility of peer pressure to carry on sniffing, for example, using video role play, or exercises in how to respond to non-use of solvents. In these circumstances, the worker is the best judge in preventing an older client who is still abusing solvents having close contact with an impressionable young person.

For some clients, working in a group situation can be quite intimidating. In addition to their solvent problem, they may have difficulty in relating to people, and it would be pointless, indeed potentially harmful, for a 'blanket' mixing of clients.

One suitable, tried and tested method is where on admission the client is given a member of staff who will remain his/her main worker. Not only is this therapeutic for the client, particularly if there are other problems present (such as anxiety, depression, feelings of inadequacy) with which the personal worker can deal, but it enables the latter to gauge the client's progress; and most importantly, the client can build a non-threatening relationship with the worker that allows trust and the facility of being able to talk to a person who will listen.

If the client is no longer abusing solvents, there is then another problem: what can take the place of solvent abuse? Whatever the reasons were that started the client sniffing have to be dealt with.

Often the young person needs an older, experienced person, mother or father figure, with whom he/she can build a warm relationship, someone who will listen and offer advice when needed. Some may need help with relationships, not just in the family; family work may be needed (see p. 65): that is, involving not only the immediate family, but teachers, friends, anyone who has a direct bearing on the client's life.

HISTORY QUESTIONNAIRE

The worker should remember that there are different categories of solvent abuse; therefore before any treatment programme commences care must be taken to establish the background and history of the client's solvent abuse.

One method of obtaining relevant details quite painlessly is to have a checklist of the pertinent questions needing answers. Obviously the worker must decide when it is appropriate to use this questionnaire — trust has to be built up — but from the treatment aspect, a complete history must be obtained as soon as possible.

The questionnaire can include the following, divided into sections on family, school, social and personal:

Family

Number of members in the family?
Any problems within the family?
How did parents react when they discovered solvent-taking?

Any other members of the family have problems with solvent/alcohol/drugs?

School

Attendance problems (truancy)?
Enjoy school?
Problems at school?
Difficulty with any work?

Social

Any interests/hobbies?
Sport?
Have many friends (what age group)?
Any problems with the police?
Any outstanding charges with the police?

Personal

Type of substance?
How is it taken?
How often?
Where taken?
Date first started.
Reasons for starting.
Taken alone/with friends?
Any alcohol/drugs?
Hallucinations?
Feel any physical/mental changes since abusing solvents?

This type of questionnaire should enable the worker to ascertain what problems may initially be encountered, and will give an indication of how serious or not is the problem of solvent-taking.

INDIVIDUAL COUNSELLING

A number of books and courses are available that specifically deal with counselling techniques. However, good basic communicative skills, coupled with a knowledge of the subject, the background of professional training, or being able to call on advice from other colleagues if problems are encountered, are sufficient. Certainly there will be cases where a professional, such as social worker, nurse, or psychologist, will be needed, but if the team is multidisciplinary-based, advice, help, and experience can be shared to the team's benefit.

The main points for individual counselling can be summarised briefly:

(a) a close relationship with one key worker where trust and understanding can exist;

(b) to be able to listen to the client, allow him time to talk about his problems;

(c) if he is attempting to abstain and fails, be positive about when he was abstaining, discuss the reasons why he went back onto solvents, and try to find a pattern;

(d) encourage participation of the family;

(e) encourage outside interests as substitutes for solvents;

(f) goals can be set, for example, attendance at school; or contracts can be used, for example, abstinence from one week to the next. Positive reinforcement of the contract will give the young person a goal to work for.

Difficulties can arise if the client continues to abuse solvents; the general behaviour pattern deteriorates, and one is met with complete indifference to the problem, so other strategies must be used.

GROUP WORK (TALKING, PROBLEM-SOLVING AND SHARING)

To reinforce abstinence from solvents, and help with personal problems, allow a youngster to observe that he/she is not the only person with certain problems. Ex-solvent abusers can act as an example, showing others that it can be done and being able to share their experiences. One word of caution: group work in this context can be potentially very therapeutic, but it can also go the other way and actually end up reinforcing solvent-taking. Experienced staff should always be present.

The basic rules for group work with young people are the same as for adults in a group setting:

(a) There must be a quiet *group room*, enough space for a small number to sit comfortably, and close to each other, in a circle that enables everyone to observe and be observed.

(b) The *therapist* will be present, not only to observe but to participate — not too actively, otherwise the group will look upon him as the major 'problem-solver' and use him to make decisions when these should be effected through group interaction. Young people do not like being patronised. The therapist should try to achieve an equal balance in his approach without making any in the group feel inadequate, in a context that is acceptable to both parties; with some thought given to vocabulary, communication can be established with ease.

(c) Care should be taken when choosing *members of a group*. For example, if the majority of the group are relatively young, and group sessions have shown positive support with each other's problems, possible confrontations and negative sessions can occur if an older user who is still actively involved in solvents is admitted. Here, even with intervention from the therapist, the new member, being older, will have some appeal to the younger (and generally immature) members of the group. To prevent this type of incident occurring, one solution is individual counselling, until it is felt that the user would benefit more from group work and interaction with other young people with similar problems.

(d) The *type of group* needs to be chosen with care: some young people work well when there are mixed members (of both sexes), others where there is a large age range. Some can be more responsive with a group occupation and others will find that verbal interchange is suitable for them. The choice of therapist also extends the range: male, female, different ages.

Generally most members can adjust to their group. It would be an impossible task to create the ideal group situation for every person. It is therapeutic for all members of a group to experience minor adjustment in a controlled setting, and it can be a starting point for members to support each other.

In the setting of group work for young abusers, there are two main 'group work' situations: closed and open.

A closed group consists of a small group of members who meet regularly; the membership never changes, a time limit is set as to length of each meeting (say, one or two hours), and each 'course' (say, six weeks), usually at the formation of the group.

An open group works almost in reverse: ever-changing members, no length of time, but with a time limit for each meeting. This type of group is to be favoured. There will be a demand for places, and in conjunction with other therapies, it can offer the user a positive approach to his problems.

To allow for a cohesive treatment programme, and a realistic group work approach, we can expand the uses of the group. Not only can it be a focal point for discussion, but various occupations can be included, all of which help to promote group and individual interaction and to allow the less verbal to communicate in other ways.

Other options can include the following.

Newspaper quizzes

This is a very simple exercise; every person is given a copy of a newspaper, with ten minutes for reading it, followed by a quiz on its contents. This can be either team-orientated or individual. This type of activity encourages conversation and some competitiveness. It also gives a bonus in the form of encouraging reading and an awareness of topical happenings.

Health education package

Basically the group is asked to design a package aimed at school-children to prevent solvent abuse. It is hoped, through discussion, that the many problems associated with solvent abuse can be talked about and, using the 'package' as an aid, can offer clients the opportunity to discuss what should be included in it and why. This exercise can sometimes lead to a greater depth of understanding their own problems as it can be easier for them to talk about solvents while using the package as a focal point.

81

These and other similar activities can be spaced out, and if timing is good, can prove to be of benefit to group members, improving relationships, interaction and support amongst the group.

Note: While a talking group is in full swing and solvent-taking is the subject being discussed, it is important that members who have cut their intake or stopped taking solvents should be allowed to share their experiences with the others and to explain what has helped them to abstain. They must be encouraged and supported. A healthy balance should be maintained between the newer members who still may be sniffing and other group members who have managed to stop. This will assist in fostering discussion and communication between members. Time and therapeutic value can be lost if long discussions are permitted on the best substances and ways of getting high. It can be extremely easy to fall into this trap, because it is a non-threatening and easy subject to discuss.

ART THERAPY

Along with other creative therapies art can be used very effectively in the treatment of young solvent abusers; it is mainly practised by occupational and art therapists.

It is adaptable and gradable with a valuable end product. Its uses can therefore range from pure activity to projective art in analytical forms which can include: painting, collage, murals, and crafts (many can now be bought in kit form, for example, marbling and enamelling).

The aims of art therapy can be described as follows.

Establishing rapport

First, it is essential to gain the trust of the client. This can be aided by any activity, not purely art although in this instance the age group concerned may relate more readily to art than to some other activities.

Carrying out an activity can help break down barriers sometimes created in a more formal interview situation, thereby allowing the client to relax and be less guarded and defensive. It provides a topic that can be used as a tool to lead into conversation.

By incorporating the above, valuable information can be obtained from the client in a non-threatening situation. The worker and client can work together, improving trust and giving the client a sense of achievement, thus reinforcing self-esteem and self-worth.

Concentration

Art can be used not only to assess fairly accurately the level of concentration in a client, but can also be graded to improve the level and standard of work, and allow the worker and client to see the gradual increase in ability.

Outlet for aggression and frustration

Some activities such as murals and pottery are extremely effective in facilitating these symptoms in a controlled manner, being active, strenuous therapies. Softening clay by pounding is an ideal way 'to let off steam', and an essential part of the activity. Self-expression can also help, just through creating! The client can then reassess the situation in a calmer, more controlled, and more realistic way; defensive barriers can also be broken down.

Social interaction

With the use of art as a medium, the level and appropriateness of social interaction can be assessed or improved.

Working on similar or group projects can facilitate this, as communication is essential even at a basic level; for example, asking 'can you pass the paint' provides a topic for conversation, or at least an opening. The aim is to encourage a sense of responsibility, and the ability to cooperate with others, which should lead to the forming of relationships both individually and as a group.

83

Organic problems

Potential organic damage may be assessed through specific art/ craft activities as deficiencies in perception, colour discrimination, hand/eye coordination can be observed with other symptomatology associated with transient or permanent organic damage.

Stimulation

This is essential in order to use any activity as a therapeutic medium. Without interest, and especially motivation, it can be an uphill struggle to carry out the activity. But if it is stimulating it should reinforce and encourage motivation and further interest, allowing the client to expand interests which can be shared with others.

Thematic art

This is an analytical projective type of art therapy, mainly practised by art therapists, and promoting exploration and expression of emotion. A theme is given to the painting/ drawing, and the end product is then explained by the client and discussed.

Some clients may find it difficult to recognise, let alone discuss, problems, particularly those associated with trauma and emotion.

Art in this context can be an ideal medium for expression and realisation of problems. Through discussion and exploration it can facilitate expression and acceptance of emotions thus allowing the client to re-establish primary needs.

Note, however, that projective art therapy should only be carried out by a trained professional in this field. Over the past few years, art as a medium for treatment has become more popular and is being recognised as a positive therapeutic activity.

Art work, therefore, can provide the potential development of interests and leisure activities as a positive alternative to solvent abuse, and encourages the promotion of self-esteem and confidence.

Precautions and contraindications

One should be aware of the safety factors involved in any activity, especially art and crafts.

On a basic level it should be checked as far as possible whether any clients have an allergy to materials used.

Great care should be taken with any substances that can be used for sniffing, such as varnish, felt pens, etc. (it is impossible to remove potential substances for abuse entirely from clients' lives); but using substances in a constructive fashion can allow them to see and participate in activities where these objects are used as originally intended. Certainly there can be potential hazards in using materials that can be abused, but non-toxic substances are available. Working in a well-ventilated area should prevent any residual fumes.

Whenever possible, use good quality equipment; this will promote motivation far more than using cheap, poor quality equipment.

Any activity should not only be appropriate, but well planned, with all workers aware of how to carry it out, to avoid failure by the client.

The client's own worker will be aware of benefit from art/ craft therapy, and it will lend a degree of trust to the situation; once the client realises that he is being put on trust, hopefully he will not abuse any of the substances.

Obviously, not all solvent abusers will benefit from this treatment. Care must be exercised when clients are being assessed for particular activities.

OCCUPATIONAL THERAPY

The occupational therapist, because of her training, can bring a number of therapeutic activities to bear upon the intervention and treatment of solvent abusers, including individual counselling, education and the use of social skills.

The occupational therapist can observe and assess clients, identify their needs and implement a treatment programme with other workers, to meet the client's needs. It is essential that there is effective communication within the treatment team.

Generally, the occupational therapist involved with the treatment of solvent abusers is attached to a treatment centre and is part of a multidisciplinary team.

The main aims that can be achieved using occupational therapy would be:

(a) improve concentration;
(b) encourage a positive attitude to increasing self-esteem and self-confidence;
(c) increase insight into the effects of solvent abuse;
(d) encourage expression of emotions;
(e) increase motivation;
(f) improve self-responsibility;
(g) encourage decision-making and problem-solving;
(h) improve appropriate social interaction and communication.

SOCIAL SKILLS

If utilised sensibly, social skills can form part of a constructive treatment programme, as not only do they encourage insight into behaviour, but clients can also be helped to change some aspects of themselves. In order to use social skills properly with the client (or diagnostic group), it is important to establish both (a) trust; and (b) an effective peer group.

Clients will be far more compliant and motivated if advice comes from the peer group, not the professional, and group members should be made responsible for this. Both 'warm-up' and 'trust' exercises are used, and these can lead on to role-play.

Warm-up exercises

The members sit in a circle in a comfortable room. They are given some instructions: introduce yourself, and then give two positive factors about yourself (it has to be beneficial, something you believe is positive or true to yourself).

Everyone then introduces himself/herself to the person on their right-hand side and states two positive factors that they believe of that person, who will reply back with his/her name and two positive factors. This continues until all members have spoken equally, followed by a discussion about peoples' feelings regarding the positive statements already made. There are many variations on this theme using different topics.

Trust exercises

Break the main group into smaller groups of approximately six people; five of the group members stand in a circle with the sixth inside it. This person should allow him/herself to be passed from person to person (standing straight with arms by the sides and eyes closed).

Another variation is where all members of the group clasp their hands and gently moving their bodies back while keeping the feet in one position (a wave-like effect), support each member in turn (who keeps a straight back). Yet another variation is where half the group close their eyes, and are led around the room by the remainder of the group who then swap round.

These exercises are useful for stimulating conversations and communication regarding feelings and emotions.

Aims of social skills

(a) to gain insight into one's own behaviour;
(b) to encourage ability in making self-decisions and to accept responsibility for actions;
(c) to identify any problem areas in one's own skills;
(d) to practise situations that are threatening or stressful;
(e) to find alternative approaches if the present one is not effective;
(f) to learn how to recognise non-verbal communication;
(g) to realise the effect certain behaviours have on others;
(h) to learn from others;
(i) to plan for both short-term and long-term goals.

Psychological help

If the client has profound difficulties, and if staff are in any way doubtful about their client's prognosis or feel unsure about the level of problems that he/she is experiencing, further professional assistance, from a psychologist for example, would be appropriate. Psychodrama is a very strong therapeutic tool, but it should only be attempted with the assistance of a fully qualified therapist.

USE OF COMPUTERS

Computers are now part of our everyday life. Young people in particular use computers both at school and at home. By using a computer as a therapeutic and recreational activity, some clients can achieve abstinence from solvents where other more orthodox methods have failed. Computers, however, are merely the tool in this situation, as it is peer group pressure that stops solvent abuse.

Several years ago a computer was donated to the Acorn Street Clinic in Glasgow, where its initial use was to be for recreation. However, after a few months, some noticeable trends became apparent. If the computer user was still taking solvents, he would generally achieve low scores, demonstrate poor concentration, and have decreased reaction time. But for those who had stopped abusing solvents, scores increased (using different games), concentration improved, and reaction time became faster.

From these findings it was decided to use the computer on a more therapeutic basis. By having a league table with the highest scores at the top, it became a matter of pride to achieve high scores on the computer games, thus introducing a competitive element into the treatment. As the young people competed with each other (not the staff), it appeared to become a point of honour with them not to abuse solvents which would decrease their scores, and make them lose face amongst their non-abusing peers.

The most popular computer games appear to be space games, racing cars, and football, both sexes finding these games appealing. Some chronic abusers have found this activity to be extremely revealing to themselves; they have noticed a definite decline in their actions, particularly if younger clients are forging ahead with their scores.

VIDEO ROLE-PLAY

This has been used with a small group, of mixed sex and age, with whom a bond of trust has developed between staff and themselves. Trust is vital as role-play can provide a degree of trauma. The principles of role-play are explained to them, and at this point they have the opportunity to examine the equipment and to experiment taking short films of each other. The main theme of this role-play is the 'effects of a member of the family who abuses solvents'.

Once it has been established who will be playing the family members, the remaining clients are encouraged to assist with the filming. Obviously time and patience are needed here, and there will be various unscheduled breaks due to laughter, shyness and perhaps uncertainty as to their particular role.

When a short film has been shot successfully, the work must be maintained therapeutically. While filming, the young people role-playing family members can be too busy acting to realise the implications of their roles.

After viewing the film, discussions can involve:

(a) resolution of crisis;
(b) parent/child relationships and relationships with other members of the family.

Young people sometimes unconsciously view their problem through 'actors' to achieve a solution, thus observing the problem in a different way, so staff involved must be aware of this, and introduce a reality element into their discussions.

HYPNOTHERAPY

Solvent abusers, like other drug addicts, usually have personal, emotional and social difficulties that are intricately linked with their behaviour. They are often described as more tense, excitable and emotionally volatile than their non-abusing peers. Also they have a markedly poor self-concept, and lack persistence, willpower and social and personal effectiveness. These characteristics, together with less emotional maturity and tolerance of frustration, provide aetiological cues to the behaviour of solvent abusers. There is a close relationship between inner drives and outward behaviour.

Further, the role of environment, ranging from accidental chance happenings to the individual's perception and understanding of his environment and his difficulties in dealing with social reality, cannot be ignored. Studies have shown that solvent abusers usually have a high potential intelligence, though academically they are poor achievers. They also have a difficulty in dealing with their environment personally, constructively and objectively, leading to a lack of motivation and boredom. Indulgence in solvent abuse can be seen as a reaction to their personal and social inadequacies and their attempts to compensate for them.

In view of the above, hypnotherapy was included in the treatment strategy for its obvious advantages. Its main aim was to help those children with training in deep mental and physical relaxation to reduce and later overcome their physical and mental tensions. In the initial assessment for suitability for hypnotherapy, 40 per cent of the children who attended the Clinic were found to be suitable and were given this treatment. The rest were considered unsuitable due to a variety of reasons including antisocial behaviour. It should be stressed that for hypnotherapy a careful assessment is essential. It is well known and documented that hypnotherapy may well enhance antisocial behaviour. Those unsuitable for hypnotherapy received a combination of therapies mentioned above in this book (pp. 76–89).

The suitable children were first given four to five half-hour weekly sessions in developing hypnorelaxation, followed by ego-strengthening techniques. In later sessions, it helped them to divert their attention from glue-sniffing to other constructive and desirable activities. They were training to break and dismiss their craving for glue-sniffing, with the help of controlled suggestions for posthypnotic conditioning. They were also given controlled suggestions to help them counteract social and peer group influences.

It should be noted that hypnotherapy was not used as an isolated therapeutic technique, but as an integral part of a therapy package that included social, occupational, diversional, computer, drama and so on.

The use of imaginal aversion techniques as a part of hypnotherapy was discussed with all those who received this treatment; 10 per cent of the children agreed to develop a dislike and distaste for the solvent. All were helped to develop a distaste for the foul smell whenever an attempt was made to sniff solvent.

Most of the other children developed their own individual methods for disliking solvents, and avoiding situations in which they could be vulnerable and resist peer group pressures. Of those who received hypnotherapy the majority showed a marked improvement in anxiety tension levels, quality of sleep and rapid decline in their habit. However, in some cases where hypnotherapy alone was used, little improvement was shown in behaviour, except some reduction in the frequency of abuse.

Hypnotherapy then is a useful technique to help solvent abusers, but only if it is used in conjunction with other techniques which encompass all aspects of behaviour, personal as well as emotional and social.

10

Setting up a Specialised Treatment Centre

Establishing a specialised treatment centre for solvent abusers can be achieved without much difficulty, provided a few basic guidelines are followed. There are several equally important areas to work through.

MANAGEMENT

Before the treatment centre opens, decisions must be made regarding the management structure, both internally and externally. A management committee can be formed consisting of representation from the centre, along with the relevant management members, who would probably be various heads of departments from the different agencies working there. The treatment centre may be staffed by members of one profession or, ideally, a mixture of different professionals from all disciplines, and may be run full-time, part-time, evenings or weekends.

Internal management would be in the hands of a project leader who would assume responsibility for the day-to-day running, and would be accountable to the management committee. Some centres have a management committee structure, but also rely on an executive committee for policy decisions. It is essential that all team members are actively involved in policy decisions.

AIMS OF A TREATMENT CENTRE

It is essential that the workers in the treatment centre have an understanding of what their aims are. This should be accepted by all of them and by whatever committee is overseeing the project. The following relevant points have a direct bearing on the success of the treatment centre:

(a) times of opening;
(b) type of abuser: will other types of abuse be permitted, for example, drugs, alcohol?
(c) treatment programmes;
(d) age groups;
(e) accepting clients from court;
(f) advertising;
(g) finance.

Opening times

Solvent abusers are generally young people still attending school, therefore, a slight dilemma creeps in: if the clinic is being run full-time, are the youngsters motivated to attend for treatment, or is it an easy option to school? If circumstances dictate that they must attend full-time, a teacher could be provided if necessary (in a residential setting there should be adequate provision for young people not to miss their schooling), who will give the essential teaching requirements.

In a number of treatment centres these problems do not arise as they are run on a part-time basis, clients attending either in the evenings or at weekends. Working with young people on this basis can be just as beneficial as having a full-time attendance, as they are not only only being treated for their solvent problem, but can still carry out their normal daily routine, deal with everyday problems, and learn to cope with their abuse problem on a daily basis, gaining support and confidence from the centre.

Once the hours are decided, and the centre has been functioning for a period of time, take care not to change the hours (unless increasing them) as this leads to confusion and uncertainty especially if clients turn up and the centre is closed. New opening hours should be widely advertised.

93

Type of abuser

The treatment centre may be the only available one in the community, so it is reasonable to expect that not only solvent abusers, but also those abusing drugs or alcohol may be referred. A decision must be made regarding their admission: will they be able to receive treatment in the solvent abuse centre, and if the centre is the only one available, perhaps facilities can be made for other substance users? However, in an area where other agencies are available, clients can be referred. This underlines the importance of all staff being aware of all available local facilities, and how to refer clients to them.

As described earlier one cannot always categorise solvent abusers as taking solvents only, neither do drug users just take drugs — there is considerable overlap; so unless otherwise indicated it is preferable to treat all abusers in the solvent abuse centre. If staff are unsure about certain other substances, or have little knowledge about them, expert assistance can be obtained.

Treatment programmes

As previously discussed in Chapter 9, a variety of treatments are available that can be utilised in any treatment centre. Obviously staff require training to participate in these activities, although most professional workers have skills already developed from their training and experience.

Inservice training can be arranged, and experience and knowledge-sharing developed. Various agencies throughout the country run courses on pertinent areas. As with any programme concerned with substance abuse, a variety of treatments are available when appropriate. The decision as to which can be offered to clients should be decided by all staff concerned after a preliminary interview with the client (possibly at a case conference). Whenever treatment programmes are decided for a client it is advisable for at least two staff members to discuss the programme.

Age groups

What age would the treatment centre start treating clients, and would there be an upper age limit? It is not unknown for some solvent abusers to be as young as five or six years. But unless there are other problems associated with this young age group intervention and treatment at a centre can be considered too traumatic; generally any youngster of such an age will sniff solvents if they have found the remains of glue or gas left by an older abuser, or (somewhat more worrying) they may be emulating an older brother or sister. The final decision on how to intervene in this type of situation should be with the family and all relevant agencies.

Solvent abusers can also be adults in their 30s and 40s, so there should be facilities available for this age group too. Usually the older client responds best to individual counselling and support; they are generally lonely, usually unemployed, and have a low self-esteem and poor self-image. If correctly managed and motivated to stop taking solvents, they may become voluntary workers assisting at the centre because of their personal experience, and they can relate and identify with younger clients. Their contribution can be very helpful, particularly with chronic abusers who in the main will listen to a person who has had a similar problem. If this is not feasible, voluntary work with an agency can be arranged.

Accepting clients from court

As mentioned in Chapter 4, treatment centres for solvent abusers are now receiving an increasing number of court referrals to help young people brought before them on charges related to solvent abuse. Invariably they are either put on probation, or given a deferred sentence providing that they attend the centre regularly for treatment.

Advertising

Without any form of advertising the facilities at the treatment centre will not be utilised fully. It is important that agencies having contact with the centre, no matter how slight, are fully aware of its role, how referrals are made, treatments available, who to contact, and opening hours. Posters with details can be distributed to general practitioners' surgeries, casualty wards, youth clubs, and colleges; small advertisements can be placed in local newspapers.

One useful method of advertising is to give short talks on the services being provided to local groups, rotary clubs, women's guilds, or teachers' associations.

Finance

No organisation can exist without some form of finance. For the purposes of this chapter, it will be presumed that the relevant agencies are providing necessary finance for the building, equipment, running costs and salaries of a specialised treatment centre. Chapter 11 gives a more detailed account of how finance can be raised.

At this stage other pertinent policies should be studied and discussed:

    (a) referral system;
    (b) catchment area;
    (c) internal management;
    (d) training of staff;
    (e) support;
    (f) outside agencies.

REFERRAL SYSTEM

Most facilities insist on some form of referral system, perhaps a letter from a general practitioner, or from one professional agency to another. However, as we are dealing with a young age group, there should be encouragement for young people to attend treatment centres with the minimum of inconvenience. An open referral system is more practical, and ensures that young people are not given a roundabout referral system — some young solvent abusers finding it difficult to ask one agency to refer them to another for help. The open referral system also means that anybody walking in off the street for help can be seen — parents, relatives, friends, as well as professional agencies can all refer cases. A telephone call or letter should be sufficient to alert the centre about a new client. Appointments, when appropriate, should be given as soon as possible, and adequate staff should be available for self-referral arrivals. These should be made welcome, shown around the premises, have treatment explained, introduced to the staff and any other clients who are already attending.

CATCHMENT AREA

As a rule, hospitals and various local authority services have catchment areas (if one lives in the south of a city, for example, that area's services are available to you, and generally, unless there is a special reason unavailable if you live outside that particular area). However, as not many treatment centres are available for solvent abusers, it would be a sensible policy if catchment boundaries were waived. Some clients, for various reasons, prefer to attend for treatment in centres not in their own locale.

INTERNAL MANAGEMENT

Roles in the treatment centre — who does what — must be clearly established. Hopefully a specialised setting like this would have a multidisciplinary approach. A project leader from any discipline would make daily management decisions. To alleviate potential stress or problems from working in this environment, regular staff support meetings should be held, in addition to any clinical or case-history sessions.

TRAINING OF STAFF

This consists of two interconnected areas:

(a) training in gaining knowledge of what solvent (and other substances) abuse consists of, with its associated problems;
(b) training in how to participate in treatment programmes.

Learning about solvents

Professional staff wishing to work in this type of specialised clinic, although they may have other skills, often lack knowledge of the subject. Various activities can be arranged to increase staff's awareness of solvent abuse, inservice training can be introduced, visits to other treatment centres can be arranged, and staff could possibly be seconded for a period. This is particularly useful as it will allow for all aspects of treatment given in an existing clinic to be experienced by the future worker, and problems previously encountered by the existing clinic can be discussed and hopefully avoided when the new centre opens.

Various organisations do send out information on solvents on request and libraries can obtain articles published on the relevant subject. Staff from established centres can hold seminars or discussions on solvent abuse.

Treatment programmes

Training of staff for treatment programmes can be extremely varied; obviously treatments vary from establishment to establishment although the basic principles apply:

(a) to provide the means to allow the client to abstain from solvents;
(b) to promote clients' physical and psychological wellbeing;
(c) to prevent relapse by support and guidance.

Sample histories can be given to give staff experience of various cases, allow them to explore treatment strategies, the use of relevant agencies, and give them the opportunity to discuss the case with experienced staff.

Two sample case-histories are given below, both different but demonstrating the need for outside agencies to be involved, and a decision made regarding suitable treatment.

Case 1: Mary
Mary, aged 14, lives in a children's residential home, and has a history of abusing a variety of solvents for two years.

Background. Mother is divorced and remarried, and living on the other side of the country; there are now two siblings. Mary has no contact with either mother or father (father's whereabouts are unknown). She was referred to the clinic from the children's home.

Management. The case looks complicated, but taken step by step will allow staff to observe how other agencies function, and the value of case conferences. Variations on the programme can be taken to help this young girl, but a treatment strategy can be formulated using these basic guidelines:

(a) medical history, physical examination, take blood for liver function tests;
(b) discussion between residential and treatment staff over continuity of care plan;
(c) care plan, individual counselling, group work, increase of social skills, have an ex-user, possibly a girl the same age, to work with her and offer encouragement;
(d) possibility of fostering, or contact with mother;
(e) weekend job;
(f) ensuring stability.

Case 2: Peter

Peter is 21, and a glue abuser for over five years, and has been involved with the police on several occasions for breach of the peace, assault, and theft.

Background. He lives at home with his aged mother; brother and two sisters, both married, live away from home, father is dead. At present Peter is awaiting sentence for minor charges. He has been given a deferred sentence in order to obtain help for his glue-sniffing from a treatment centre. Social background reports have been requested.

Management. The following guidelines can be used:

(a) medical history, physical examination, take blood for liver function tests;
(b) obtain all relevant reports: social work, court and probation reports;
(c) case conference with all staff concerned;

99

(d) supportive group work and individual counselling;
(e) finance, ensure that both the mother and Peter receive correct social security payments;
(f) check problems at home: mother perhaps disabled; any alcohol/drug abuse;
(g) voluntary work available.

Taking up the last point: if treatment has proved successful, perhaps Peter can assist at the treatment centre as a volunteer for a time, or be encouraged to participate in other organisations, providing support for the handicapped or similar. Using these facilities to allow ex-solvent abusers the opportunity to assist with the less able in society has met with considerable success. Obviously great care must be taken in ensuring that the ex-abuser has the correct motivation for this work. Taking on voluntary work provides him not only with a purpose and a sense of achievement, but allows him to gain self-respect and build his self-confidence.

SUPPORT

In any organisation there are some administrative duties. In a treatment centre, there should be back-up staff to deal with paperwork, answering phones, typing reports, etc. This staff plays an important role, and should be encouraged to be part of the team; they should have at least a rudimentary knowledge, not only about solvent abuse, but how to cope with the wide spectrum of cases, and sometimes alarmed, distraught parents.

Clear, concise records should be maintained, and records of attendance, especially if dealing with court referrals, should always be up to date. Over the years useful research material can be gleaned from records such as suitability of treatments, how clients have responded to the centre's regime, and how successful intervention has been.

Ideally, sufficient administrative staff should be available to allow the clinical team to be involved solely with clients, thus maximising their clinical skills.

OUTSIDE AGENCIES

As mentioned before in this chapter, it cannot be stressed how important these agencies can be. Clients will be referred to the clinic who have had, or are presently receiving, support and help from other agencies, so a great deal of background material can be obtained.

If no medical practitioner is attached to the centre, local general practitioners may wish to offer their services for physical examinations.

On discharge, the client may require extra support from various agencies who can assist in the resettlement of such cases, and a variety of other ways.

11

Support and Help
from the Voluntary Sector

INTRODUCTION

It is gratifying to note that volunteer groups have shown an increasing interest in setting up centres for helping young solvent abusers; hopefully they will work in conjunction with professional bodies, who will support them in all areas.

Initiating a treatment centre requires motivation, enthusiasm, and a very strong commitment. Voluntary centres can be started for a variety of reasons — by an ex-abuser wishing to help others, the parents of a child abusing solvents, or well-meaning members of the public who feel that they would like to contribute their time and energy towards helping young people with a solvent problem.

If an individual wants to establish a project for solvent abusers, he will need other helpers. This can be solved by approaching an existing voluntary group, or centre, perhaps through a church group or an advert in the local paper.

Once a small group of volunteers has been established several important aspects have to be taken into consideration. The aims and objectives should be carefully examined and made as realistic as possible.

(a) finance;
(b) premises;
(c) leadership, supervision, management.

FINANCE

Costs will be incurred for either part-time or full-time centres. If the centre is run under the auspices of a professional body, funds can possibly be channelled into the centre; however, if this is not the case, the individual must obtain funds for the costs of such a project — finance may be required for the purchase or rent of suitable premises, for example.

If the total costs are not too prohibitive, financial support can be solicited from local businesses, or from general fund-raising. Suitable premises — for example, a local church, or social work department — may be made available.

Urban aid programmes

Fortunately, funds are available through the Urban Programme (Urban Aid) for such projects, and armed with the appropriate information groups who are interested in the running of such centres can apply for funds.

'Urban Aid was initiated in 1968 in response to a growing concern by central government about social deprivation in urban areas. Legislation on this — the Local Government Grants (Social Need) Act 1969 — is an example of a policy that allocates additional resources to areas of special need. (The Act authorises the Secretary of State for the Environment to pay grants to local authorities with urban areas of special social need.)

In Scotland the Regional Council no longer receives an allocation of Urban Aid from the Scottish Office for new projects but rather submits each year a programme of new and supplementary applications which are assessed on a competitive basis with submission from other regional and district councils.

The urban renewal unit has recently been transferred from the Scottish Development Department to the Industry Department for Scotland.

Urban programme grants are now for an initial period of 4 years and good projects run by voluntary organisations and community groups may be extended by a further 3 years.

In England preference is usually given to capital projects, or to those where the provision of a few salaried workers is sufficient to mobilise the resources of the community itself.

The urban programme is not intended to provide permanent funding for voluntary projects because this would restrict its ability to support new schemes. It is a mark of the valuable and worthwhile nature of voluntary schemes that very few cease to exist when their urban programme funding time expires. Most schemes not reapproved are absorbed into local authority main programmes. Of the small number of voluntary schemes that lose both urban and main programme funding, most have usually either achieved what they set out to do, or found other sources of funding.

Essential criteria for the urban programme should be understood, as follows:

(a) *Target group:* the project should be directed towards a local or national priority client group — for example, the under-5s, the elderly, ethnic minorities, 'latchkey' children, or *young people in danger of abusing drugs or otherwise at risk, or a combination of such target groups.*
(b) *Target area:* the project should be defined to meet the needs of an identified urban area; it is not enough that inner city residents would incidentally benefit from it.
(c) *Innovation:* the urban programme is well suited to support pilot and experimental projects, ways of tackling growing social problems, and help develop new patterns of service provision, including projects run by voluntary or other groups which embody a new approach to problems, or fill gaps in existing services.' (Department of Environment, 'Urban programme').

When considering making application for a grant through the urban programme, up-to-date criteria and information regarding current practices in awarding aid should be obtained. The information supplied above is only intended as a guideline for sources where funds can be made available.

PREMISES

Finding suitable premises which are inexpensive and situated where potential clients can attend without much difficulty is possibly the most difficult part. The building should be large enough to handle clients, and have sufficient separate rooms for individual and group activities running concurrently.

If the centre is to be run on a part-time basis, obtaining premises may not entail so much difficulty as local schools, church halls, and social work departments may be willing to assist.

However, if the centre is to be open full-time, the building would have to be purchased purely for the purpose. If available finance is minimal, help may come from agencies involved in other aspects of drug/alcohol dependency by either sharing their own premises, or allowing them to share the new centre's facilities.

Once the location is established it must have adequate security. A sad fact of life is that treatment centres attract an undesirable element who attempt to break in thinking that syringes or drugs are kept there.

An ideal layout for the centre would be two large rooms suitable for a variety of activities, three smaller rooms for individual treatments, and a bright, cheerful admission area. Provision must be made for stationery, filing cabinets, and possibly a small first aid facility.

LEADERSHIP, SUPERVISION, MANAGEMENT

Any organisation requires some form of management, and a volunteer centre is no different.

There are various ways to achieve a management structure. However, although it is a relatively simple task setting up this structure, if it proves to be unwieldy and cumbersome with different committees making decisions, it is difficult then to reduce these structures to a more realistic decision-making process.

Part-time administration

For part-time administration a management committee should be made up of members of the volunteer group and perhaps representation from the local health authority and social services department. The committee would be responsible for the centre's aims and policy-making, defining roles, programme planning, and training. Hours of opening, referral systems and catchment areas are covered fully in Chapter 10. From this view there are few differences between a specialised treatment centre run by professionals and a volunteer group.

An advisory or executive committee may also be recommended, the members consisting of various heads of department (from local agencies), and a representative from the financial source, if relevant. Generally these committee members have a wide experience of the management and function of a project dealing with various problems, and as such can prove invaluable for advice and assistance. The management committee should be accountable for the general running of the centre, and while the centre is open there should always be a member present.

As decisions have to be made daily, it would be impracticable to expect all members of the management committee to be in attendance while the centre is open, so a project leader responsible to the committee should be in attendance regularly. He/she should have a deputy to ensure that there is always somebody present who can make decisions for the centre.

Full-time administration

With the centre functioning full-time, the main difficulties that emerge will be an increase in funding and sufficient staff to provide a comprehensive programme.

Once the main management structure is in force, other aspects should be examined to ensure that the centre is working at its fullest potential and that all members are aware of their roles.

The most important points to be discussed would be:

(a) monitoring;
(b) roles;
(c) staff meetings;
(d) induction of new members.

Monitoring

This would consist of collecting information on the work being carried out at the centre, reviewing the service to see if the correct criteria are being met, attendance of sufficient numbers, staff awareness of their function, or any other serious problems.

The monitoring role could be carried out by the project

leader, and any findings reported to the relevant committee after discussion with the centre members. The main function of monitoring is to prevent any problems escalating into more serious issues, and will provide the staff with a realistic method of ensuring that the centre's role is being met.

Roles

Difficulties can arise unless there is a clear policy on roles in the centre. To prevent staff discontentment there should be a clear definition of who does what. Lists and rota systems can be drawn up detailing work and sharing of all activities, including the least popular ones.

If this is not suitable, members can put forward their own particular expertise, and if this allows all the activities to be covered, from administration to treating clients, this method should then be used.

Staff meetings

These allow for support and give staff the opportunity to discuss any problems. They should be organised on a regular basis, with minutes and an agenda. Staff should feel free to mention any topics of their work, and should expect support and guidance from other members.

Induction of new members

An induction programme would assist new volunteers to settle in quickly, and begin to participate in all activities, particularly for the first few sessions at the centre. A more experienced staff member should always be with them until it is felt that the new member has sufficient expertise and knowledge to deal with clients without supervision.

TRAINING

All members of a voluntary group have something to offer, whether time or a special skill. Initial meetings should be used to formulate what skills are available to be shared with other members. These meetings can also be used to discover what training is needed, and who can provide it.

Local authorities, social and community workers, and health professionals all have relevant experiences, and assistance can also be gained from existing voluntary groups.

Nobody expects voluntary groups to be running acute, dynamic therapies for solvent abusers, but they are providing a badly needed service which can complement professionally run centres, and with good basic skills training, they are able to offer valuable help to solvent abusers and their families.

Good contact is essential between the voluntary sector and the professional clinic, who will be able to help each other in many ways.

Unless any member of the voluntary centre has particular expertise in the treatment of solvent abusers, it would be advisable to abstain from using the more complex forms of treatment such as group psychotherapy (intensive), hypnosis, projective art work (see Chapter 9). This should in no way detract volunteers from extending their knowledge in other aspects of care, one of the most important being to provide an alternative to solvent sniffing. Some centres put great emphasis on recreational retraining, and this can be achieved by an introduction to activities in the centre which, with encouragement, can be practised outside the centre. These can include: computer activities, model making (generally the more complex and sophisticated the better), table tennis, snooker, swimming, cookery classes, making clothes.

Having a listener to hear the client's problems and offer advice is also very important. The worker will often find that the young person can have difficulty talking to teachers or doctors, but once a relationship has started to develop between them, perhaps interspersed with some games, the client will have built up some trust and will feel secure enough in the relationship to talk about his solvent problem and any other difficulties he may be facing.

Being able to talk out his problem in a non-threatening situation, the client can begin to examine his lifestyle, and with suitable encouragement, will raise his self-esteem and gain confidence. Group meetings, for which staff do not require too much training, can discuss a variety of topics, and also act as a medium to encourage non-use of solvents by ex-users.

WORKSHOPS

Workshops can be set up with the assistance of local health

education and social work staff where staff can explore the problems associated with solvents, and gain knowledge of suitable strategies to use themselves when dealing with clients and their families.

Information can be obtained from libraries or other functioning clinics, and various organisations send out literature upon request.

Appendix A

Mortality Statistics[1]

Difficulty can often be encountered when trying to obtain information about what is essentially an extremely complicated piece of research. To actually obtain information in such depth to provide us with valuable insight into the varieties of death and substance causing it should hopefully spur on those of us who are involved in the care and treatment of solvent abusers to try and reduce the number of young people who are becoming statistics (see also Anderson *et al.*, 1982).

It may be of interest to note that the final comment in Anderson *et al.*'s (1986) paper states

> . . . our data were collected using consistent methods over five years. Thus, though knowledge of, and interest in, the problem has grown, there has probably been a real increase in the number of deaths. This contradicts the belief that the abuse of volatile substances is a passing fashion. Our evidence indicates that the various efforts which have been made to prevent abuse and reduce the risk of death among abusers have not affected the scale of the problem, the pattern of substances abused, or the risk of accidental death among those abusing volatile substances.

DEATHS FROM ABUSE OF VOLATILE SUBSTANCES: A NATIONAL EPIDEMIOLOGICAL STUDY

Methods

Death associated with abuse of a volatile substance was defined

110

as one caused either directly or indirectly by one of these agents and not occurring at the workplace. The first source of ascertainment was through press clipping agencies employed by three different organisations to scan all newspapers in the United Kingdom for articles containing specified key words relating to the subject: 'aerosol' (British Aerosol Manufacturers Association), 'glue-sniffing' (Institute for the Study of Drug Dependence), and 'sniffing deaths' (Anderson, McNair and Ramsey, 1985; Anderson et al., 1986).

The second source of ascertainment was a list of names compiled non-systematically from miscellaneous sources from 1970 to 1980 initially by the Department of Trade and then by the Department of Health and Social Security.

The third main source was a regular systematic survey of coroners in England, Wales, and Northern Ireland carried out at six-month intervals from 1982. Deaths were also notified to us from the Office of Population Censuses and Surveys. In Scotland we were aided by the Crown Office and the procurators fiscal. After the detection of a death associated with abuse of a volatile substance, the inquest proceedings, postmortem and toxicological findings, and death certificates were obtained. The strength of evidence connecting the death with abuse of a volatile substance was graded sequentially as: toxicological findings positive for volatile substances; subject witnessed to be sniffing at time of death; circumstantial evidence in a known 'sniffer'; and circumstantial evidence without a history of sniffing. Cause of death was classified in such a way as to help distinguish:

(1) those subjects in whom death was almost certainly not directly caused by the volatile substance (for example, trauma);

(2) those in whom the direct role of the volatile substance was questionable (plastic bag found over the head, aspiration of stomach contents);

(3) those in whom the only explanation for the death was that it had been a consequence of the direct toxic effects of the volatile substance itself.

Statistical significance was calculated by the chi-square test. Confidence intervals for standardised mortality ratios were calculated by two methods: if the observed number of deaths was 30 or less a table of the exact confidence interval for the

111

mean of a Poisson variable was used; when the observed number was greater than 30 the standard error formula $\sqrt{O/E}$ was used (where O and E represent observed and expected numbers of deaths, respectively).

Results

Between 1971 and 1983 the number of deaths detected increased year by year, reaching 80 in 1983 (Table A.1). The increase was found by each method of ascertainment, including the period 1981–83, when methods were applied prospectively. In most cases (80 per cent) there was some mention of volatile substances on the death certificate, though not necessarily among the stated causes of death.

Table A.1: Deaths Associated with Volatile Substance Abuse 1971–83; Numbers, Sources of Ascertainment, and Mention of Volatile Substance on Death Certificate.

	Year													Total
	71	72	73	74	75	76	77	78	79	80	81	82	83	
Number of Deaths	2	4	3	2	9	7	7	14	20	29	45	60	80	282
Source of ascertainment—press clipping (key word "aerosol") (since 1970)			2	1	4	2	4	3	5	10	8	6	9	54
Press clipping (key words "glue sniffing") (since 1976)					2	3	1	4	2	18	23	28	54	135
Press clipping (keywords "sniffing death") (since 1981)		1					1	1	2	14	39	39	65	162
Coroner's survey. Scottish Crown Office. (Since 1982)								1	7	7	3	31	45	94
Departments of Trade and Health (ceased 1980)	2	4	3	2	6	6	3	7	9	2				44
Death certificate obtained (number with mention of solvents)	2 (0)	3 (0)	3 (0)	2 (0)	8 (6)	7 (6)	7 (6)	13 (11)	17 (15)	27 (22)	45 (32)	59 (52)	65* (64)	258 (208)

* Some outstanding due to normal delay.

The following data were obtained for the deaths: inquest proceedings (73 per cent of cases); necropsy results (94 per example, letters from coroners (30 per cent). The strength of evidence connecting a volatile substance with the death was, in decreasing level of certainty: toxicological findings positive for volatile substances (67 per cent of cases); witnessed to be sniffing at time of death (6 per cent); circumstantial in a known sniffer (8 per cent); circumstantial only (14 per cent); not known (4 per cent).

Figure A.1: Age Distribution in Deaths Associated with Volatile Substance Abuse in the United Kingdom, 1971–83.

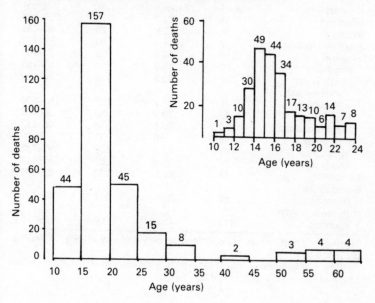

Figure A.1 gives the age distribution of the subjects at death. The median age at death was 17 years, with a range of 11–76 years; 44 subjects (16 per cent) were under the age of 15, and 201 (71 per cent) were under 20. Most (95 per cent) were male. Analysis by the time periods 1971–80, 1981–82, and 1983 showed no evidence of a trend in age distribution, and there was no significant association between age distribution and the method of ascertainment.

Using the total for 1982 as the denominator, abuse of volatile substances in 1983 in Great Britain was estimated to account for

1.5 per cent of all deaths in boys aged 10–14, 2.4 per cent in youths aged 15–19, and 0.6 per cent in men aged 20–24. As a proportion of deaths due to injury and poisoning the respective figures were 3.2, 3.4, and 0.9 per cent.

Table A.2: Deaths Associated with Volatile Substance Abuse 1971–83, by Country and English Region

Country/region	Number of deaths	Expected	Age standardised mortality ratio	95 Per cent confidence interval
Scotland	48	27.0	178	127–228
Wales	11	13.8	80	40–143
Northern Ireland	7	8.4	83	33–172
England	216	232.7	92	80–105
Northern	16	15.9	101	58–163
North West	38	33.3	114	78–150
Yorkshire and Humberside	21	24.7	85	53–130
East Midlands	13	19.2	68	36–116
West Midlands	20	26.5	76	46–117
East Anglia	3	9.0	33	7–97
South-east	91	83.3	109	87–132
South-west	14	20.8	67	37–113
London	44	33.31	132	93–171
Metropolitan counties	60	58.64	102	76–128
Non-metropolitan counties	112	140.80	80	65–94

Testing null hypothesis of no difference between countries $\chi^2 = 18.3$; 3 df; $p < 0.001$.
Testing null hypothesis of no difference between types of county within England: $\chi^2 = 8.78$; 2df; $p < 0.05$.

Table A.2 shows the geographical distribution of the deaths. There was a highly significant variation in standardised mortality ratios in the United Kingdom, with the highest in Scotland (178) and the lowest in Wales (80). Within England the variations were more difficult to interpret because of the random variation associated with small numbers.

There was, however, no evidence of a north–south gradient. The possible effects of environment were examined by comparing the standardised mortality ratios of counties within England. This showed a significant association, the standardised mortality ratio being highest in London (132), intermediate in metropolitan counties (102), and lowest in non-metropolitan counties (80). There was no evidence of a shift in the regional distribution of deaths over time; there was, however, a highly significant trend towards younger deaths in the north (χ^2 test for trend 10.3; $p < 0.01$).

Table A.3: Deaths Associated with Volatile Substance Abuse, by Social Class*

Social class	Age < 16 years (head of household's occupation)		Age ≥ 16 years (own occupation or if not known or student, head of household's)		Total‡		Census 1981 (head of household)
	Number	Per cent	Number	Per cent	Number	Per cent	Per cent
I	4	5	9	5	13	5	6
II	11	15	26	15	37	15	24
III Non-manual	6	11	15	9	21	9	12
III Manual	27	36	45	26	72	29	34
IV	17	23	31	18	48	19	16
V	9	12	30	17	39	16	5
Armed forces	1	1	16	9	17	7	3
TOTAL	75	100	172	100	247	100	100

* Insufficient information on 35 deaths.
‡ Goodness of fit using 1981 Census to estimate expected deaths: $\chi^2_6 = 77.4$. $p < 0.001$.

The distribution of deaths by social class was significantly different from that described for the whole population at the 1981 Census (OPCS, 1983), being more common than expected in social class V and the armed forces (Table A.3); nevertheless, all social classes were well represented, and the overall impression was that the association with social class was small. Of subjects aged under 18, 10 per cent (17) were in care compared with an expected 0.8 per cent. Of those aged 18 or more, 72 per cent were unmarried and 67 per cent were employed or students. Deaths were most likely to occur at home, usually in the bedroom (118 cases; 42 per cent), or in a public place (81; 29 per cent). A further 30 subjects (11 per cent) died in hospital, though some of these were almost certainly dead on arrival. There was no significant association between deaths from abuse of volatile substances and month of the year. There was, however, a significant association with the day of the week ($p < 0.01$), Saturday showing the highest number and Thursday the lowest.

In most cases (71 per cent) the victim had been sniffing while

115

alone. In 17 (8 per cent) of those for whom inquest data were available there was evidence that the inhalation of solvents was associated with autoerotic behaviour. Among this group, constriction of the neck with or without a plastic bag over the head was observed in six cases and some form of bondage in

Table A.4: Volatile Substances Abused, by Product and Age

Product	AGE (years)				Total (per cent)
	<16	16–17	18–24*	≥25	
		Gas fuels			
Lighter fuel	19	16	11	1	47
Other products	1	0	4	3	8
Not known	7	2	3	1	13
Total number of deaths	27	18	18	5	68(24)
		Aerosol sprays			
Pain relief (PR) spray	5	7	3	2	17
Fire extinguisher	5	4	1	–	10
Antiperspirant	5	1	–	–	6
Air freshener	1	1	2	–	4
Cleaners	1	–	1	2	4
Fly spray	1	–	–	1	2
Paint spray	–	1	3	–	4
Hair spray	–	–	–	1	1
Not known	–	–	1	–	1
Total number of deaths	18	14	11	6	49(17)
		Solvents in glues			
Named brand with toluene	12	20	13	6	51
Glue (with toluene)	2	3	5	2	12
Glue (non-toluene)	2	–	4	–	6
Not known	2	3	2	1	8
Total number of deaths	18	26	24	9	77(27)
		Other volatile substances			
Cleaning agents	9	4	8	7	28
Plaster remover	6	5	5	–	16
Correcting fluid thinner	5	5	–	–	10
Glitter lamp fluid	3	–	–	–	3
Chloroform	1	–	1	3	5
Petrol	1	1	–	–	2
Paint Thinner	–	1	2	–	3
Other	–	–	2	1	3
Not known	5	4	3	5	17
Total number of deaths	30	20	21	16	87(31)
Grand total of deaths	93	78	74	36	281(100)

*One solvent not known in 18–24 year age group.

four. The most common verdict was misadventure (63 per cent), followed by accident (27 per cent), open verdict (4 per cent), and suicide (5 per cent). There was no overall association between verdict and age, though with increasing age there was a significant trend away from a verdict of misadventure towards an open or suicide verdict (χ^2 for trend 5.84; $p < 0.01$).

The commercial product was not always recorded. When more than one substance was involved we usually accepted the coroner's opinion on which was the most important. Table A.4 lists the main substances implicated in the deaths. Solvents in glues were associated with just over a quarter of deaths. None of this group had any mention of a substance from one of the other groups. Eight different brands were mentioned, the most frequent being Evostik (74 per cent). Of the 52 cases with a toxicological report, 42 showed evidence of toluene, followed by trichloroethylene (3), hexane (3), 1,1,1-trichloroethane (2), tetrachloroethylene (1), and methylethylketone (1).

Other volatile substances were implicated in 31 per cent of deaths and included at least 20 different products. Two cases in this group had mention of a substance from another group (butane, 1; aerosol spray, 1). Toxicological analysis in 79 cases in the group detected 16 different volatile substances, the most frequent being 1,1,1-trichloroethane (42 cases) followed by trichloroethylene (10), chloroform (7), carbon tetrachloride (5) and toluene (4). One quarter of deaths were associated with the abuse of gas fuels containing the alkane gas butane or, less commonly, propane. Gas cigarette lighter fillers accounted for 80 per cent of this group.

Aerosols were the least common category (17 per cent). These usually contain halocarbon propellants 11 or 12, but may include some butane. Other groups mentioned in the aerosol group were solvents in glue (2 cases), gas fuels (2), and in two cases the aerosol delivered a solvent. At least 13 products were incriminated but the most common was pain relief (PR) spray, followed by fire extinguishing agents containing bromochlorodifluoromethane.

Non-volatile potentially harmful or contributory substances were occasionally detected toxicologically. The most common were alcohol (24 cases), followed by barbiturates (4), amitriptyline (1), and benzamphetamine (1). These were not significantly associated with the category of volatile substance.

There was no significant association between the category of volatile substance and the verdict or the strength of evidence connecting the substance with death. The relative importance of each category of volatile substance did not change significantly over time, nor were there any associations with age or with country or region.

Table A.5 analyses the different mechanisms or cause of death by broad category of substance. Trauma accounted for 31 deaths (11 per cent), of which 11 were due to drowning, nine to hanging, and 11 to multiple or head injuries. Traumatic deaths accounted for a much higher proportion of deaths associated with glue than other substances. In the 58 subjects (21 per cent) who were found with a plastic bag over the head, and the 50 (18 per cent) who showed evidence of inhaled stomach contents, it was not certain whether death occurred for these reasons or was caused primarily by direct toxic effects. There remained 142 deaths (51 per cent) which could be assumed to have resulted directly from the toxic effects of the substance itself.

There was no evidence of a change in the relative distribution of cause of death over time, nor was there an association with country or region of death. There was evidence of a trend towards plastic bag-associated deaths with increasing age, but this failed to reach significance ($\chi^2 = 3.36$; $p = 0.07$).

RECENT TRENDS IN MORTALITY ASSOCIATED WITH ABUSE OF VOLATILE SUBSTANCES IN THE UNITED KINGDOM

A report was made in 1985 on 282 deaths associated with abuse of volatile substances that occurred between 1971 and 1983. Trends were difficult to determine because the methods of investigation had not been consistent. The same methods, however, were used to detect deaths from abuse of volatile substances over the past five years and trends can now be examined.

Attempts have been made in recent years to curb the abuse of volatile substances, including legislation and voluntary codes of practice concerning the sale and display of certain products, and the dissemination of information to professionals, parents, and children. Data were examined for the five years 1981–85 to identify the current trends and determine whether recent control measures have had any effect.

Table A.5: Mechanism of Death, by Substance

Mechanism of death	Substance									
	Gas fuels*		Aerosol sprays		Solvents in glues		Other volatile substances		Total	
	Number	Per cent	Number	Per cent	Number	Per cent	Number	Per cent	Number	Per cent
Trauma‡	0		1	2	29	38	0		30	11
Plastic bag over head	15	22	10	20	21	27	12	14	58	21
Inhalation of stomach contents	15	22	10	20	4	5	21	24	50	18
Direct toxic effects of substance	37	55	28	57	23	30	54	62	142	51
All mechanisms	67	100	49	100	77	100	87	100	280	100

* In one death cause not known.
‡ In one death substance not known.
$\chi^2_9 = 95.1$; $p < 0.001$.

Methods and results

The methods of investigation, described above, rely mainly on a systematic survey of newspapers by a press clipping agency; regular surveys of coroners; and liaison with the Office of Population Censuses and Surveys, and in Scotland the Crown Office. Inquest proceedings, necropsy reports, and toxicology findings are sought in all cases and obtained in most. Emphasis was on trends in numbers nationally and regionally, age and sex distribution, the various substances abused, and the cause of death. Significance was measured using the chi-square test.

There were 385 deaths during the five years, the yearly number increasing from 46 in 1981 to 116 in 1985. Regional trends varied, and a comparison of the numbers for the two years 1982 and 1983 and those for 1984 and 1985 shows that deaths occurring in Wales, Yorkshire and Humberside, and West Midlands regions doubled, while there were smaller increases in Scotland and Northern Ireland, and no increase in the South-east. Most of those who died (285; 74 per cent) were under the age of 20, and 65 (17 per cent) were aged 10–14. No

Table A.6: Number (per cent) of Deaths Associated with Abuse of Various Volatile Substances in the United Kingdom 1981–85

	1981	1982	1983	1984	1985	TOTAL
Gas fuels	15(33)	11(18)	19(24)	31(38)	30(26)	106(28)
Aerosol sprays	1(2)	8(13)	12(15)	9(11)	20(17)	50(13)
Solvents in glue	16(35)	18(29)	24(30)	15(19)	35(30)	108(28)
Other volatile substances	13(28)	24(39)	25(31)	26(32)	28(24)	116(30)
Substances unknown	1(2)	1(2)	0	0	3(3)	5(1)
Total	46	62	80	81	116	385

trend was found in the age or sex distribution of deaths, and there was no change during the period in the proportion of deaths associated with the different substances, with an increase in numbers occurring in all major categories (Table A.6). Overall, gas fuels (mainly butane), solvents in glues (mainly toluene), and 'other solvents' (principally plaster removers and correcting fluid thinners, mainly 1,1,1-trichloroethane) each accounted for around 30 per cent of deaths.

In 53 per cent of cases death was attributed to the direct toxic effects of the substance. The remaining deaths were thought to have resulted from intoxicated behaviour (trauma, 15 per cent), the method of inhalation (plastic bag over the head, 16 per cent), or inhalation of stomach contents (16 per cent). There was a significant downward trend ($p < 0.01$) in deaths associated with plastic bags from around 20 to 10 per cent; otherwise no trend in the cause of death was observed.

NOTES

1. I am extremely grateful to the authors and the *British Medical Journal* for allowing me to use the substance of Anderson, McNair and Ramsey (1985) and Anderson *et al.* (1986) to demonstrate the mortality aspects in the abuse of volatile substances and Ms K. Bloor for her kind help and assistance.

Appendix B

Teaching Aids[1]

TEACHING AIDS

Free to Choose (TN)[2]

Dealing with Solvent Misuse (TN)

Both are obtainable from Teachers Advisory Council on Alcohol and Drug Education (TACADE), 2 Mount Street, Manchester M2 5NG. Telephone 061 834 7210.

Drugs Demystified Training Pack (TN)

Contains a full set of guidelines and materials for running short participative inservice training courses on drug, alcohol and solvent-related problems. Particularly suitable for multi-disciplinary training. Obtainable from Institute for the Study of Drug Dependence, 1–4 Hatton Place, Hatton Garden, London, EC1N 8ND.

Facts and Feelings about Drugs but Decisions about Situations (TN)

Teacher's manual with units on facts about alcohol, drugs, solvent abuse, etc. and on decision-making skills about situations in which drugs may be offered. Fact sheets and flashcards to be reproduced and distributed to class. This is a short 'situational education' course for secondary schools obtainable from ISDD (as above). ISDD Research and Development Unit, 1982.

Health Careers (TN)

Teacher's manual of 13 units (lesson guides) covering drugs,

122

alcohol and solvents, in the context of the transition from school to work and associated youth cultures. A full school-leavers' course integrating work education and health education. Obtainable from ISDD (as above). Dorn, N. (ISDD Research and Development Unit) and Nortoft, B., London: ISDD, 1982.

Wread (Work Related Education on Alcohol and Drugs (TN)

The pack provides methods and materials to make the trainer effective in helping young people develop into adults who avoid unwise use of alcohol, drugs and solvents. It also helps them to improve communication, problem-solving and social skills. Obtainable from ISDD (as above).

Solvent Abuse — Who Sniffs What, When and How (TN)

An information pack for youth workers, teachers, and other adults working with young people. Obtainable from Chief Executive, Merseyside Youth Association, 88 Sheil Road, Liverpool, L6 3FA. Telephone 051 263 0557.

Young Men's Christian Association (TN:V)

The YMCA has 200 local centres throughout the country. An investigation into drug and solvent misuse has led to a developing programme and new resources, including a video. Obtainable from National Council of YMCAs, 640 Forest Road, London E17 3DZ.

Solvent Abuse Teaching Pack (TN:V)

Obtainable from Department of Education, Robertson Teachers Centre, 16 Glasgow Road, Paisley, PA1 3QG.

Illusions (V:TN)

The film should be used as part of a seminar or training course for professionals on the subject of solvent abuse. A comprehensive set of notes has been compiled to assist the course organiser. This presenter's guide is in two parts. The first part is intended to help trainers and teachers use the film to greatest advantage, and it offers questions and scenarios which can be used as basis for discussion; it also elaborates on the strategies for intervention illustrated in the film. The second part is intended to give background information for the students themselves and should be copied for each individual. A full bibliography is also included. Prepared by the Central Office of

Information. Obtainable from Central Film Library, Hire and Sales, Chalfont Grove, Gerrards Cross, Bucks. Telephone 02407 4433.

Solvent Abuse (V)

A video by Kent Police TV Unit for showing to children by teachers and other professionals. Available from Kent County Constabulary, Police TV Unit, Force Training School, Coverdale Avenue, Maidstone, Kent ME15 9DW.

Take a Deep Breath (V)

Teaches shop staff about sniffing and how to deal with it. This is a video made by Woolworth for staff training and available to all retailers without restriction.

Not to Be Sniffed At (V)

The growing and serious problem of solvent abuse, often referred to as 'glue-sniffing', was the subject of this 25 minute documentary film in the award winning BBC 'Scene' series, which is aimed at the 14–16-year-old school audience, and provided by the BBC at the request of the School Broadcasting Council for the United Kingdom. An excellent video for general viewing. Obtainable from BBC Enterprises Film and Video Library, Guild House, Oundle Road, Peterborough, PE2 9PZ. Telephone 0733 63122.

My Life (V)

Elain Patterson, a glue sniffer, interviewed by Colin Morris on Yorkshire TV. Obtainable from Non-Theatrical Programmes, Yorkshire Television, Leeds LS3 1JS. Telephone 0532 438283 (also available as individual episodes).

In a Different World — Glue Sniffing (V)

This 30 minute 16 mm or video film tells of a 20-year-old's fight against his addiction to glue-sniffing which wrecked his life and ended with him setting fire to his flat, leaving him homeless and destitute. Available for hire from Concord Films Council, 201 Felixstowe Road, Ipswich, Suffolk IP3 9BJ. Telephone 0473 715754.

On the Glue (V)

This 30 minute 16 mm or video film by Thames TV investigates the dangerous and sometimes fatal effects of glue-sniffing among children. Available for hire from Concord Films Council, 201 Felixstowe Road, Ipswich, Suffolk IP3 9BJ. Telephone 0473 715754. The video may be purchased from Thames Television International Limited, 149 Tottenham Court Road, London W19 9LL. Telephone 01 387 9494.

Kid's Stuff (V:C)

A film produced by Project Icarus with accompanying audio cassette on the history and implications of solvent abuse and the life and death of a sniffer. Obtainable from Project Icarus, 4 Clarence Parade, Southsea, Portsmouth PO5 3NU. Telephone 0705 827460.

British Medical Television No. 29 (V)

An 'on the beat' policeman talks with knowledge and experience about solvent abuse in a revealing way (filmed on location in Basingstoke). Dr Joyce Watson, Vice President of Re-Solv, discusses the medical aspects and in particular the general practitioner's role. Available at £6.00 per copy inclusive of postage and packing from BMTV Limited, 3 and 4 Woking Business Park, Albert Drive, Woking, Surrey GU21 5JY.

Solvent Abuse, by the Health Education Council (S:TN)

This is a 63 slide presentation and training manual for professionals providing 10 sessions of about 90 minutes each. It can be viewed at the Health Education Council, 78 New Oxford Street, London, WC1A 1AH. Telephone 01 637 1881. (Packs obtainable from: Michael Benn and Associates Limited, PO Box 5, Wetherby, Yorkshire, LS23 7EH. Telephone 0937 844524.

Glue Sniffing (Volatile Substance Abuse) (S:C)

A two part slide/tape presentation: Part 1 — History of Solvent Abuse, part 2 — Signs and Symptoms of Solvent Abuse. Obtainable from Camera Talks Limited, 31 North Row, Park Lane, London W1R 2EN. Telephone 01 493 2761.

Re-Solv: Solvent Abuse Advisory Kit (for retailers only)

Includes leaflets of explanatory warning door/shelf stickers. Produced by Re-Solv, The Society for the Prevention of Solvent Abuse, St Mary's Chambers, 19 Station Road, Stone, Staffs.,

APPENDICES

ST15 8JP. Telephone 0785 817885.

Volatile Substance Abuse Information Pack

Information sheets dealing with stereotypes, indicators of use, methods of use, substances which are abused, first aid. For people working in statutory or voluntary agencies. Available from Hereford and Worcester Social Services Department, County Hall, Spetchley Road, Worcester.

Lifeline Project

Lifeline produces printed and audiovisual training material on drugs and solvent misuse and possible response in practice. It also runs the Regional Drug Training Unit based at Prestwich Hospital. Lifeline is located at Joddrell Street, Manchester M3 3HE.

LEAFLETS AND PAMPHLETS

Inclusion in this listing does not necessarily imply suitability: we suggest that you check for yourself.

Adfer Unit

West 1 Ward, Whitchurch Hospital, Cardiff, CF4 7XB
The Adfer Unit treats people with drugs, alcohol and solvent problems and helps in rehabilitation. It has produced many publications.

City of Birmingham District Council

Education Welfare Service, Room 213, Education Department, Margaret Street, Birmingham, B3 3BU
Glue-Sniffing

Department of Health and Social Security

Dept DM, DHSS Leaflet Unit, PO Box 21, Stanmore, Middlesex, HA7 1AY
Drug Misuse: A Basic Briefing

Dewsbury Community Health Council

The Town Hall, Dewsbury, West Yorkshire, WF12 8DG
Solvent Abuse — A Guide for the Professional

126

Evode Limited (for retailers only)

Common Road, Stafford, ST16 3EH
Glue Sniffing — An Important Message

Health Education Council

78 New Oxford Street, London WC1A 1AH
What to Do About Glue Sniffing

The Health Education Service and Avon and Somerset Police Juvenile Bureau
Solvent Abuse and Your Child

Hope Press Publications

45 Great Peter Street, London SW1P 3TL
Sniffing It — Snuffing It
Stuck with It
Solvent Abuse in the Family
Trend Magazine Issue 1

Kick-It

National Office, 6 Church Street, Wolverton, Milton Keynes
MK2 5JN
Don't Sniff It — Kick It!

Lifeline Project

Joddrell Street, Manchester, M3 3HE
Sniffing for Pleasure by Rowdy Yates
Guidance notes on counselling young solvent abusers.

P. Linaird and Co (Publishing)

Unit 5, 61–63 Brownfields, Welwyn Garden City, Herts.
Glue Sniffing — Information for Parents

Lions Club International

Available from your local Lions Club, or Mr R. Everest,
'Vardecott', Cairn Road, Ilfracombe, North Devon
Is your Child Stuck on Glue Sniffing?

Merseyside Drugs Council

25 Hope Street, Liverpool 1
Glue Sniffing — How to Cope
Why Sniff?

National Campaign Against Solvent Abuse

245a Coldharbour Lane, London SW9
Advice Sheet for Parents
Advice Sheet for Professionals
More Than Just a Passing Phase
Solvent Abuse in School
Answers to Questions on Solvent Abuse (for the user)

National Confederation of Parent–Teacher Associations

43 Stonebridge Road, Northfleet, Gravesend, Kent DA11 9DS
Solvent Abuse: Does it Concern Us?

Release Publications Limited

1 Elgin Avenue, London W9 3PR
Sniffing Glue and Other Solvents

Rev. Paul T. Arnold

KTS Hall, 50 Gold Street, Kettering NN16 8JB
A Look at Solvent Abuse

Sandwell Health Authority

Kingston House, High Street, West Bromwich
The Glue-sniffing Problem

South Wales Association for the Prevention of Addiction

111 Cowbridge Road East, Cardiff CF1 9AG
Solvent Abuse and Glue-sniffing — notes for parents, teachers
and those concerned with young people

Strathclyde Regional Council

82 West Regent Street, Glasgow G2 2QF
Sniffing — The Facts

The Wirrall Committee on Solvent Abuse

WCVS 222 Liscad Road, Wallasey L44 5TN
Solvent Misuse

Voluntary Service, Aberdeen, Solvent Abuse Project
4 Albyn Place, Aberdeen AB9 1RY
Solvent Abuse — Why?

NOTES

1. The information contained in this appendix has been kindly supplied by Re-Solv (The Society for the Prevention of Solvent and Volatile Substance Abuse), St Mary's Chambers, 19 Station Road, Stone, Staffs ST15 8PJ.
2. TN: teaching notes; V: video; C: audio cassette; S: slides 35 mm.

Appendix C

Relevant Government Departments

This information, which is kindly supplied by the Home Office in their summary of the government's strategy regarding 'Tackling Drug Misuse', explains which government departments are responsible for the various aspects of drug misuse, including treatment.

NOTE ON DEPARTMENTAL RESPONSIBILITIES

The Home Office is responsible in England and Wales (and, in certain respects, Scotland) for the scope, effectiveness, enforcement and administration of the controls in the Misuse of Drugs Act 1971 and its associated subordinate legislation. It has an overall responsibility for coordinating the development and implementation of the government's policies and a specific coordinating responsibility in relation to prevention policy.

The Department of Health and Social Security (DHSS) is concerned with services for the treatment and rehabilitation of drug misusers and with health education (other than in relation to schools, etc.) in England.

The Department of Education and Science (DES) is concerned with policy at the national level in England, health education in schools, other maintained educational institutions and the youth service. The provision of health education in these institutions is a matter for local education authorities and the institutions themselves.

The Board of HM Customs and Excise is responsible for the prevention and detection of the illegal import and export of drugs.

The Foreign and Commonwealth Office assists in determining policy towards the international bodies and other governments. Its overseas posts act as a channel of communication and its staff attend meetings when attendance of an official from the United Kingdom would not be justified.

The Overseas Development Administration provides financial and technical assistance to developing countries, including some producer and transit countries.

Two departments of the Scottish Office, the *Scottish Home and Health Department* and the *Scottish Education Department*, have responsibilities which broadly correspond to those of the Home Office, DHSS and DES in England. SHHD is principally responsible for the enforcement of the controls in the Misuse of Drugs Act 1971 and its associated subordinate legislation, for the National Health Service and services for the treatment and rehabilitation of drug misusers, for health education in the community and for police matters. SED is responsible for health education in relation to schools and other educational institutions, for the youth and community education services and for social work matters.

The Welsh Office is concerned with services for the treatment and rehabilitation of drug misusers and with health education in Wales.

The Department of Health and Social Services, Northern Ireland, is responsible for the provision of services for the treatment and rehabilitation of drug misusers and for health education (other than in relation to schools etc.) as well as for enforcement and administration of the controls in the Misuse of Drugs Act 1971. It also has a general coordinating responsibility in relation to Northern Ireland interests.

The Department of Education, Northern Ireland, is responsible for health education in relation to schools, other educational institutions and the youth service.

Appendix D

Statistics on 600 Solvent Abusers

The following statistical information on 600 solvent abusers who attended the Acorn Street Solvent Abuse Clinic, Glasgow, has been compiled from a short questionnaire (figures are in percentages). This information is only intended as an approximate guide to various trends noticeable at the Solvent Abuse Clinic. Further detailed information can be obtained from the author.

(1)	*General breakdown of referrals*		
	Drug Advisory Service (locally based project)		30
	Teachers (school nurses)		20
	General practitioners		20
	Health visitors and district nurses		15
	Self-referrals		10
	Others (friends, ex-attenders, relatives, etc.)		5
(2)	*Age*		
	5–10		4
	11–15		57
	16–20		26
21–25		7	
	26–30		3
31–35		1	
	36–40		0
	41–45		2
(3)	*Sex*		
	Male		70
	Female		30

(4)   *Size of family*
      (includes parent(s) and all siblings)

2	5
3	14
4	19
5	19
6	14
7	14
8	7
9	5
11	3

(5)   *Marital status of solvent abuser's parents*

Married	53
Separated	18
Divorced	16
Widowed (one parent)	11
Died (both parents)	2

(6)   *Housing*
      (no catchment area for the clinic;
      situated in the centre of Glasgow)

Local authority housing (council)	80
Private housing	20

(7)   *Criminal record*
      (including breach of the peace, car theft,
      assault, stealing, attempted murder, arson,
      drug trafficking

Criminal record	45
No criminal record	55

(8)   *Type of solvent*

Glue	38
Aerosol	13
Butane gas	32
Cleaning agents	4
Petrol	2
Correcting fluid thinner	6
Magic mushrooms	2
Pain relief (PR) spray	3

In addition, 26 per cent of the above were involved in drug/ solvent mixing.

Butane gas/aerosol + cannabis	10

Butane gas/aerosol + LSD	2
Butane gas/aerosol + amphetamines	8
Butane gas/aerosol + Temgesic	2
Butane gas/aerosol + DF118 (or similar)	3
Butane gas/aerosol + barbiturates	1[1]

No data on glue and drug mixing.

(9) *Sniffing situation*

Group	66
Sniffing alone	34

Approximately 14 per cent of the total number have taken solvents in both situations.

(10) *How often solvents are taken*

Daily	27
Four/five times weekly	21
Two/three times weekly	31
Once weekly	8
Once only	13

NOTE

[1] The above figures are as accurate as possible.

Appendix E

Some Useful Addresses

The majority of addresses have been kindly supplied by Re-Solv (Society for the Prevention of Solvent and Volatile Substance Abuse).

National Campaign Against Solvent Abuse
Avon Co-odinator: Mr R. Hancock
13 Downlands Road
Fishponds
Bristol

Bristol Campaign Against Solvent Abuse
Mrs M Johnson
60 Charnwood Road
Whitchurch
Bristol
Telephone 0272 834138

Kick It
National Office
6 Church Street
Wolverton
Milton Keynes MK2 5JN
Telephone 0908 368869 or 0924 477146
(branches nationwide)

Mrs S Cherry
Essex Campaign against Solvent Abuse
42 Charles Pell Road
Colchester
Essex
Telephone 0206 869058

East Dorset Drugs Advisory Service
79 Old Christchurch Road
Bournemouth BH1 1EW
Telephone 0202 28891

Dr L Lowenstein
Consultative Committee on Solvent Abuse
Allington Manor
Allington Lane
Fair Oak
Hants SO5 7DE
Telephone 0703 69262

Isle of Wight Youth Trust
1 St John's Place
Newport
Isle of Wight PO30 1LH
Telephone 0983 529569

Bexley Solvent Advisory Panel
1a Pickford Road
Bexley Heath
Kent DA7 3AT
Telephone 02304 6524

Solvent Abuse Helpline
Wesley Halls
Shroffold Road
Downham
Bromley
Kent
Telephone 01 698 4415

Mrs L Hancock
Preston Campaign Against Solvent Abuse
3 Chatburn Road
Ribbleton
Preston
Lancs

Mr J Sutherland
Coordinator
Solvent Abuse Helpline
Euston Road Secondary Centre
Station Road
Morecambe LA4 5JL

Mrs G Bryce/Mrs M Sterrick
The Solvent Abuse Resource Centre
including Sniffers Anon.
Blackburn Community Block
99 Rowan Street
Blackburn
Lancs
Telephone 0506 634035

Mr Richard Ives
National Children's Bureau
8 Wakeley Street
London EC1V 7QE
Telephone 01 278 9441

Release Publications Ltd
1 Elgin Avenue
London W9 3PR
Telephone 01 289 1123

Greenwich Solvent Abuse Panel
Telephone 01 853 8110

Mr Barrie Liss
Re-Solv
St Mary's Chambers
19 Station Road
Stone
Staffs ST16 8JP
Telephone 0785 817885

Harrow Group Against the Misuse of Solvents
Telephone 01 864 6099

Mr Alan Billington
National Campaign Against Solvent Abuse
PO Box 513
245a Coldharbour Lane
London SW9 8RR
Telephone 01 733 7330

Institute for the Study of Drug Dependence
1–4 Hatton Place
Hatton Garden
London EC1N 8ND
Telephone 01 430 1991

The Lifeline Project
Joddrell Street
Manchester M3 3HE
Telephone 061 832 6353

Wirral Borough Council
Wirral Committee on Solvent Abuse
c/o WCVS
22 Liscard Road
Wallasey
Wirral L44 5TN
Telephone 051 630 4164
(Contact Mr R S Charles)

Substance Abuse Unit
'Crossways'
Whitehall Road
Uxbridge
Middlesex
Telephone 0895 57285

Mr David Hunter
East Belfast YMCA
183 Albertbridge Road
Belfast BT5 4PS
Telephone 0232 59768

Addiction Information and Resource Unit
The Information Officer

Strathclyde Regional Council
82 West Regent Street
Glasgow G2 2QF
Telephone 041 332 0062

Greater Glasgow Health Board
Solvent Abuse Clinic
23 Acorn Street
Glasgow G40 4AN
Telephone 041 556 4789 or 041 554 0405

Wester Hailes Hotline Project
The Harbour
Hailsland Road
Edinburgh
Telephone 031 442 2465

Scottish Campaign Against Solvent Abuse
Mrs A Broadfoot
29 Sidney Street
Arbroath
Telephone 0241 74712

Scottish Drug Forum
(D. Liddle and W. Slavin)
266 Clyde Street
Glasgow G1 4JH
Telephone 041 221 1175

Solvent Abuse Centre
117 King Street
Telford
Shropshire
Telephone 0952 53053

Eastbourne Solvent Abuse Committee
Mr H L McKendrick
20 Pashley Road
Eastbourne
East Sussex BN20 8DU
Telephone 0323 24982

Solvent Support Group
Options
29 Wordsworth Road
Worthing
Sussex
Telephone 0903 204539

National Campaign Against Solvent Abuse
55 Wood Street
Mitcham Junction
Surrey
Telephone 01 640 2946

Jean Bell
Solvent Abuse Counsellor
YMCA
1–3 Toward Road
Sunderland SR1 2QF
Telephone 0783 654950

South Wales Association for the Prevention of Addiction
111 Cowbridge Road East
Cardiff CF1 9AG
Telephone 0222 26113

SAVES
Solvent Abuse Voluntary Enterprises
2 Blossomfield Road
Solihull B91 1LD

Drug and Solvent Advisory Service
13 Bath Street
Swindon
Wilts
Telephone 0793 610133

Yorkshire Campaign Against Solvent Abuse
Mr T Boffin
119 Thornes Road
Wakefield
West Yorkshire
Telephone 0924 374595

Glossary

TERMS USED IN SOLVENT AND DRUG ABUSE

Acid head	Regular user of LSD
Business/gear	Equipment for taking drugs, such as syringe, needle or mirror and razor for cocaine
Bum trip	Unpleasant sensation, usually describing LSD or magic mushroom experience
Chipping	Using only a small amount of heroin at irregular intervals
Clean	Not having any drugs or equipment in one's possession
Cold turkey	Used to describe 'unpleasant' symptoms when the regular drug is not available, and flu-like symptoms at the start; but it also indicates symptoms of complete withdrawal of the drug without a substitute
Cut drugs	Adulterated drugs
Deal	Drug — money transaction
Freaked out	Generally considered to be a fairly unpleasant effect of a drug
Gun	Needle
Habit	Amount of drug needed to satisfy a user's dependency
Head	Used to describe somebody taking drugs, for example, cokehead (cocaine user), or gluehead (glue sniffer)
Hooked	Drug-dependent
Joint	Cigarette, home made (with cannabis)
Junk	Heroin

Kicking	Attempting to stop taking a drug
Machine	Syringe
Mainline	Injecting into a vein
Mainliner	Person who injects himself
Make	To obtain drugs
Pusher	Person who sells drugs
Popping	Injecting drugs into the skin
Rush	The immediate pleasurable effect of taking a drug
Score	To obtain drugs
Script	Legal prescription for drugs
Tab	An absorbent substance which has LSD in it

SLANG (STREET) TERMS FOR SUBSTANCES/DRUGS

Amphetamines	Sulph, speed, uppers, truckdrivers, wake-ups, co-pilots, thrusters
Barbiturates	Bars, blockbusters, downers, barbs
Cocaine	Heaven dust, coke, snow, nose-candy, white
Cannabis	Acapulco gold, Colombian, tea, reefer, smoke, stick, weed, pot, dope, blow
Heroin	Big H, boy, brown sugar, horse, mud, smack, skag, stuff, junk
LSD	Acid, blotter acid, sunshine
Fly agaric (Amantina muscaria)	Magic mushroom
Liberty cap (Psilocybe semilanceata)	Magic tea

Aerosols	Sniff
Butane	Sniff
Glue	Bag, pot, sniff-kit

References and Further Reading

Aerosol Review (1981) Aerosol reviews. *Manufacturing Chemist*

Akerman, H.E. (1982) The constitution of adhesives, and its relationship to solvent abuse. *Human Toxicology, 1*, 223–30

Anderson, H.R., MacNair, R.S. and Ramsey, J.D. (1985) Deaths from abuse of volatile substances; a national epidemiological study. *British Medical Journal, 290*, 304–7

Anderson, H.R., Bloor, K., MacNair, R.S. and Ramsey, J. (1986) Recent trends in mortality associated with abuse of volatile substances in the UK. *British Medical Journal, 293*, 1472–3

Asquith, S. and Didcott, P. (1983) The management of solvent abuse, an exploratory study. Research report SWSG, Glasgow

Baselt, R.C. and Cravey, R.H. (1968) A fatal case involving trichloromonofluoromethane and dichlorodifluoromethane. *Journal of Forensic Science, 13*, 407–10

Bass,M. (1970) Sudden sniffing death. *Journal of the American Medical Association, 212*, 2075–9

Black, D. (1985) Sniffing solvents and volatile substance abuse. In *Misuse of Solvents*. Re-Solv Information Pack. Re-Solv, Stone, Staffs

Bozetti, L., Goldsmith, L.S. and Ungerleider, J.T. (1967) The great banana hoax. *American Journal of Psychiatry, 124*, 678–9

British Aerosol Manufacturers Association (1975) *Aerosol ABC*. BAMA, London

British Aerosol Manufacturers Association (1978) *Alternative Aerosol Propellants*. BAMA, London

British Aerosol Manufacturers Association (1980) *Cost Effectiveness of Aerosol Products*. BAMA, London

Central Statistical Office (1984) *Regional Trends*. HMSO, London, pp. 32–3

Clark, D.G. and Tinston, D.J., (1982) Acute inhalation, toxicity of some halogenated and non-halogenated hydrocarbons. *Human Toxicology, 1*, 239–47

Cohen, S. (1973) The volatile solvents. *Public Health Review, 11*, 185–214

Cunningham, S.R., Dalzell, G.W.N., McGirr, P. and Khan, H.M.

(1987) Myocardial infarction and primary ventricular fibrillation after glue-sniffing. *British Medical Journal*, *294*, 739

Department of the Environment (1985) *The Urban Programme*. DOE, London, pp. 7, 28

Department of Health and Social Security (1982) *Treatment and Rehabilitation Report of the Advisory Council on the Misuse of Drugs*, para. 4.5. HMSO, London

de Ropp, R.S. (1957) *Drugs and the Mind*. Gollancz, London

Ditton, J. and Speirits, K. (1981) *The rapid increase in heroin addiction in Glasgow during 1981*. Background paper No. 2, University of Glasgow Department of Sociology, Glasgow

Fact Sheet 7 Children's Hearings (Scottish Information Office)

Fagan, D.G. and Forest, J.B. (1977) Sudden sniffing death after inhalation of domestic lipid aerosol. *Lancet*, *ii*, 361

Flowers, N.C. and Horan, L.G. (1972) Nonanoxic aerosol arrhythmias. *Journal of the American Medical Association*, *219*, 33–7

Francis, J., Murray, V.S.G., Ruprah, M., Flanagan, R.J. and Ramsey, J.D. (1982) Suspected solvent abuse in cases referred to the Poisons Unit, Guy's Hospital. July 1980, June 1981. *Human Toxicology*, *1*, 271–80

Garriott, J. and Petty, C.S. (1980) Death from inhalant abuse: toxicological and pathological evaluation of 34 cases. *Clinical Toxicology*, *16*, 305–15

Gossop, M. (1982) *Living with Drugs*. Temple Smith, London

Haq, M.Z. and Hameli, A.Z. (1980) A death involving asphyxiation from propane inhalation. *Journal of Forensic Science*, *25*, 25–8

Herd, P.A., Lipsky, M. and Martin, H.F. (1974) Cardiovascular effects of 1,1,1-trichloroethane. *Archives of Environmental Health*, *28*, 227–33

Hansard (1980) Sir George Young in Commons adjournment debate. Columns 203–16, *21 July*

Home Office *(1984) Statistical Bulletin* 18/84, Table 9

Home Office (1985) *Tackling Drug Misuse*, paras 1.14, 1.15, 1.16, Appendix A. HMSO, London

Institute for the Study of Drug Dependence (1984) *Drug Abuse Briefing* (Drug Laws, Legal Status). ISDD, London

Isbell, H. (1959) Comparison of the reactions induced by psilocybin and LSD-25 in man. *Psychopharmacologia*, *1*, 29–38

King, G.S., Smialek, J.F. and Troutman, W.G. (1985) Sudden death in adolescents resulting from the inhalation of typewriter correction fluid. *Journal of the American Medical Association*, *253*, 1604–6

Kramer, R.A. and Pierpaoli, P. (1971) Hallucinogenic effect of propellant components of deodorant sprays. *Paediatrics*, 48, 322–3

Kringsholm, B. (1980) Sniffing associated deaths in Denmark. *Forensic Science International*, *15*, 215–25.

Laxton, M. (1985) *Solvent Abuse — The Social Work Contribution*. Re-Solv, Stone, Staffs

Mahmood, Z. (1983) Cognitive Functioning of Solvent Abuse. *Scottish Medical Journal*, *28*, 276–80

Mee, A.S. and Wright, P.L. (1980) Congestive (dilated)

cardiomyopathy in association with solvent abuse. *Journal of the Royal Society of Medicine*, *73*, 671–2

Merck Index (1960) *Merck Index of Chemicals and Drugs* (7th edition). Merck, Rahway, New Jersey

Metzner, R. (1971) Mushrooms and the mind. In B. Aaronson and H. Osmond (eds) *Psychedelics*. Hogarth Press, London

National Institute on Drug Abuse (NIDA) (1978) Voluntary inhalation of industrial solvents. C.W. Sharpe and L.T. Carroll (eds). Departments of Health, Education and Welfare Publications, Rockville, Maryland

Office of Population Censuses and Surveys (1980) *Classification of Occupations*. HMSO, London

Office of Population Censuses and Surveys Census 1981 (1983) *National Report, Great Britain. Part 1*. HMSO, London

Pearson, E.S. and Hartley, H.O. (1970) *Biometrika Tables for Statisticians* 3rd edn. Cambridge University Press, Cambridge

Poklis, A. (1976) Sudden sniffing death. *Canadian Medical Association Journal*, *115*, 208

Ramsey, A. (1987) Drugwise 12–14. A Scottish Drug Education Programme for 12–14 year olds. *Scottish Drugs Forum Bulletin*, *3*

Reinhardt, C.F., Azar, A., Maxfield, M.E., Smith, P.E. and Mullin, L.S. (1971) Cardiac arrhythmias and aerosol sniffing. *Archives of Environmental Health*, *22*, 265–79

Roberts, D.J. (1982) Abuse of aerosol products by inhalations. *Human Toxicology*, *1*, 231–8

Sax, N.I. (1979) *Dangerous Properties of Industrial Materials*, 5th edn. Van Nostrand Reinhold, London

Strathclyde Regional Council (1984) *Urban Programme — Guidelines for Voluntary and Community Groups*. SRC, Glasgow

Taylor, G.J. and Harris, W.S. (1970) Cardiac toxicity of aerosol propellants. *Journal of the American Medical Assocation*, *214*, 81–5

Wakefield, E.M. (1958) *The Observer's Book of Common Fungi*. Warne, London

Waldron, H.A. (1981) Effects of organic solvents. *British Journal of Hospital Medicine*, *26*, 645–9

Weissberg, P.L. and Green, I.D. (1979) Methyl-cellulose paint possibly causing heart failure. *British Medical Journal*, *2*, 1113–14

Wells, B. (1973) *Psychedelic Drugs: Psychological, Medical and Social Issues*. Penguin Education, Harmondsworth

Wiseman, M.N. and Banim, S. (1987) Glue-sniffer's heart. *British Medical Journal*, *294*, 739

Index